# THE CITY'S HINTERLAND

# Perspectives on Rural Policy and Planning

Series Editors:
**Andrew Gilg**
University of Exeter, UK
**Keith Hoggart**
King's College London, UK
**Henry Buller**
University of Exeter, UK
**Owen Furuseth**
University of North Carolina, USA
**Mark Lapping**
University of South Maine, USA

*Other titles in the series*

**Women in Agriculture in the Middle East**
Pnina Motzafi-Haller
ISBN 0 7546 1920 6

**Winning and Losing**
**The Changing Geography of Europe's Rural Areas**
Edited by Doris Schmied
ISBN 0 7546 4101 5

**Critical Studies in Rural Gender Issues**
Edited by Jo Little and Carol Morris
ISBN 0 7546 3517 1

**Contesting Rurality**
**Politics in the British Countryside**
Michael Woods
ISBN 0 7546 3025 0

**Women in the European Countryside**
Edited by Henry Buller and Keith Hoggart
ISBN 0 7546 3946 0

**Equity, Diversity and Interdependence**
**Reconnecting Governance and People through Authentic Dialogue**
Michael Murray and Brendan Murtagh
ISBN 0 7546 3521 X

# The City's Hinterland
Dynamism and Divergence in Europe's Peri-Urban Territories

*Edited by*

KEITH HOGGART
*King's College London, UK*

ASHGATE

© Keith Hoggart 2005

All rights reserved. No part of this publication may be reproduced, stored in a retrieval system or transmitted in any form or by any means, electronic, mechanical, photocopying, recording or otherwise without the prior permission of the publisher.

Keith Hoggart has asserted his right under the Copyright, Designs and Patents Act, 1988, to be identified as editor of this work.

Published by
Ashgate Publishing Ltd
Gower House
Croft Road
Aldershot
Hants GU11 3HR
England

Ashgate Publishing Company
Suite 420
101 Cherry Street
Burlington, VT 05401-4405
USA

Ashgate website: http://www.ashgate.com

**British Library Cataloguing in Publication Data**
The city's hinterland : dynamism and divergence in Europe's
    peri-urban territories. – (Perspectives on rural policy and
    planning)
    1. Rural-urban relations – Europe 2. Rural-urban relations –
    Economic aspects – Europe 3. Metropolitan areas – Economic
    aspects – Europe 4. Cities and towns – Europe – Growth
    5. Europe – Rural conditions
    I. Hoggart, Keith
    307.7'2'094

**Library of Congress Cataloging-in-Publication Data**
The city's hinterland : dynamism and divergence in Europe's peri-urban territories / edited by Keith Hoggart.
    p. cm. -- (Perspectives on rural policy and planning)
    Includes index.
    ISBN 0-7546-4344-1
    1. Rural-urban relations--Europe. 2. Metropolitan areas--Europe. 3. Regional planning--Europe. I. Hoggart, Keith. II. Series.

HT384.E852C58 2005
307.74'094--dc22

2005022767

ISBN-10: 0 7546 4344 1

Printed and bound in Great Britain by MPG Books Ltd. Bodmin, Cornwall.

# Contents

| | | |
|---|---|---|
| List of Figures | | vi |
| List of Tables | | vii |
| List of Contributors | | ix |
| 1 | City Hinterlands in European Space<br>*Keith Hoggart* | 1 |
| 2 | Diversity in the Rural Hinterlands of European Cities<br>*Vincent Briquel and Jean-Jacques Collicard* | 19 |
| 3 | Commuter Belt Turbulence in a Dynamic Region:<br>The Case of the Munich City-Region<br>*Claudia Kraemer* | 41 |
| 4 | Residential Growth and Economic Polarization in the French<br>Alps: The Prospects for Rural-Urban Cohesion<br>*Nathalie Bertrand and Emmanuelle George-Marcelpoil* | 69 |
| 5 | Urban Spread Effects and Rural Change in City Hinterlands:<br>The Case of Two Andalusian Cities<br>*Francisco Entrena* | 95 |
| 6 | Tensions, Strains and Patterns of Concentration in England's<br>City-Regions<br>*Steven Henderson* | 119 |
| 7 | Convergence and Divergence in European City Hinterlands:<br>A Cross-National Comparison<br>*Keith Hoggart* | 155 |
| *Index* | | *171* |

# List of Figures

| | | |
|---|---|---:|
| 2.1 | INSEE 'Urban Regions' in France, 1999 | 24 |
| 2.2 | Percentage of NUTS2 Populations in Functional Urban Areas of 100 000 or More Inhabitants | 26 |
| 2.3 | Peri-Urbanization Gradients in Upper Bavaria | 32 |
| 2.4 | Peri-Urbanization Gradients in Rhône-Alpes and Provence-Alpes-Côte d'Azur | 33 |
| 2.5 | Peri-Urban Diversity in Andalusia and Murcia | 35 |
| 3.1 | The Munich Region | 45 |
| 3.2 | The Purchasing Power in Selected Areas in Germany | 46 |
| 3.3 | Unemployment Rates in Selected German Agglomerations, 2003 | 46 |
| 3.4 | Relative Population Change for Counties in the Munich City-Region | 49 |
| 3.5 | Percentage of Population by Age Group for Different Areas in the Munich Region | 56 |
| 3.6 | Change in Farm Sizes in Bavaria, 1979-2003 | 60 |
| 4.1 | The Location of Annecy and Valence | 71 |
| 5.1 | Percentage Population Change in Andalusian Municipalities 1991-2001 | 100 |

# List of Tables

| | | |
|---|---|---|
| 1.1 | Mean Average Values for Variables Used to Classify Rural NUTS5 Areas into Journey-to-Work Zones in Eastern England | 7 |
| 2.1 | The Demographic Importance of ESPON's Main Functional Urban Areas in European Countries | 27 |
| 2.2 | Indicators for Identifying Urban, Suburban and Peri-Urban Areas in Five European Countries | 29 |
| 3.1 | Population Change in the Munich Region 1970-2002 | 47 |
| 3.2 | Population Change in Some Fast Growing Small Municipalities in the Munich City-Region | 50 |
| 4.1 | Population Characteristics of the Urban Region of Annecy | 80 |
| 4.2 | Population Characteristics of the Urban Region of Valence | 80 |
| 4.3 | Change in Land Prices in Annecy, 1997-2001 | 83 |
| 5.1 | Number of Dwellings in Monachil by Occupancy, 1986-2001 | 102 |
| 5.2 | Percentage of the Monachil Workforce in the Construction Sector, 1950-2001 | 102 |
| 5.3 | Percentage of the Monachil Workforce in Service Industries, 1950-2001 | 103 |
| 5.4 | Percentage of Workforce by Economic Sector in La Alpujarra Municipalities 1981-2001 | 108 |
| 6.1 | Percentage of Norfolk Employment by Economic Sector, 1991-2001 | 126 |

## List of Tables

| | | |
|---|---|---|
| 6.2 | Population Growth Rates for Norfolk Local Authorities, 1981-2001 | 127 |
| 6.3 | Population Growth Within South Norfolk Settlements, 1951-2001 | 128 |
| 6.4 | Population Growth Within the Norwich Policy Area, 1981-2001 | 129 |
| 6.5 | Residents, Workplaces and Daytime Populations Aged 16-74 in 2001, by Norfolk District | 130 |
| 6.6 | Patterns of Commuting to Norwich from Selected Settlements, 2001 | 131 |
| 6.7 | Average Change in Property Prices for Selected Norfolk Local Authorities, 1999-2004 | 139 |
| 6.8 | Percentage of Non-Work Trips to Norwich During the Past Month by Village Population Size | 144 |
| 7.1 | Welfare Regimes and Housing Rental Systems | 157 |

# List of Contributors

**Nathalie Bertrand**
Researcher, Cemagref, Centre of Agricultural and Environmental Engineering Research, Grenoble

**Vincent Briquel**
Researcher, Cemagref, Centre of Agricultural and Environmental Engineering Research, Grenoble

**Jean-Jacques Collicard**
Researcher, Cemagref, Centre of Agricultural and Environmental Engineering Research, Grenoble

**Francisco Entrena**
Professor of Rural Sociology and the Structure and Change of Societies, University of Granada

**Emmanuelle George-Marcelpoil**
Researcher, Cemagref, Centre of Agricultural and Environmental Engineering Research, Grenoble

**Steven Henderson**
Researcher Associate, Department of Geography, University of Reading

**Keith Hoggart**
Professor of Geography, King's College London

**Claudia Kraemer**
Researcher, Faculty of Spatial Planning, University of Dortmund

Chapter 1

# City Hinterlands in European Space

Keith Hoggart

**Introduction**

The book brings together research in four European countries on the nature and impact on rural areas of linkages with cities. Funded by the European Commission under the Quality of Life and Management of Living Resources Programme,[1] the research for this project was designed to explore a variety of activity sectors and agents in rural zones, with a view to identifying how urban pressure on rural areas produced particular mutations and generated new dynamics of rural change. This book focused on particular aspects of this wider project, with a specific focus on the consequences for employment, housing and services of intensifying rural links with cities. The impetus behind this research was a sense that insufficient attention has been devoted to understanding city hinterlands in Europe. This is an important lapse if we are to grasp positive possibilities for future change in the European space economy. As Berg and associates (1982) recognized more than 20 years ago, Europe has shifted from seeing regional development emerging from differences in sectoral change, to a situation in which new sectoral developments depend on regional specificities. Today, the importance of local/regional institutional 'thickness', of geographically grounded networks, and of 'learning regions', are widely recognized as critical to the advancement of local and regional economies (e.g. Morgan, 1997; Keeble et al., 1999; Castells, 2000). This prompts an impetus toward identifying what different geographical contexts embody that stimulates socio-economic change. Of course, before such questions are asked, a primary issue is what kind of geographical focus should attract investigative attention. For some the answer lies in administrative regions, which explains the popularity of studies exploring differences between provincial, (formal) regional or local government areas. Undoubtedly such areas are relevant for some dimensions of human activity. But when it comes to understanding the breadth of human endeavour, activity patterns tend not to be channelled by political boundaries but

---

1   The research programme was entitled 'Urban pressure on rural areas (NEWRUR): mutations and dynamics of peri-urban rural processes' (contract QLK5-CT-2000-00094), which was funded by the DG-VI component of Framework Five under Key Action 5, on 'Sustainable agriculture, fisheries and forestry and integrated development of rural areas including mountain areas'.

rather wash over them; they might irritate and try to regulate but political boundaries rarely encompass the geography of socio-economic life (e.g. UK Royal Commission on Local Government in England, 1969). More proximate to real life choices, whether personal or corporate, are interaction systems forged around city-regions. Yet, while city-centred regions dominate large tracts of the European space (Herrschel and Newman, 2002), outdated visions of city-region fortunes being driven simply by events in the core city need discarding (Berry, 1970). It might well be, as Dunford and Perrons (1994) note, that successful regional economies are centred around major cities, but this does not mean the city as such determines the success of the city-region. In an economic climate in which there are concerns about diseconomies from over-congested cities, in which 'the environment' has become a key economic attraction, with the vitality of a region depending upon quality of life considerations, the imagery that rural areas offer 'a good life', with supportive community cohesion, a lack of social conflict and (relatively) crime-free conditions, means that country areas and rural landscapes have become central to city-region futures (e.g. Keeble *et al.*, 1992; Cabus and Vanhaverbeke, 2003).

The critical message is that rural hinterlands are not mere city 'appendages' but are integral to development processes that bear on the whole of a city-region, such that if 'appropriate' development does not occur in these areas this could impact on the whole city-region. Recently, this vision has been embodied in conceptions of a future European Union, as illustrated in fairly widespread acceptance of the principles embodied in the European Spatial Development Perspective (ESDP) (Faludi and Waterhout, 2002). As explored in more detail below, the ESDP specifies that polycentric development should be promoted across Europe, in order to enhance life chances away from the continental European economic core (Commission of the European Communities, 1999). Yet this Perspective is not just about hierarchies of city-regions. It also promotes the message that only with equality in rural-urban relationships will city-regions generate the creative capacity required to heighten local capacities to compete outside the European core. What the chapters in this book seek to explore is how far current processes within city-regions are supportive of the kind of integrated development that the ESDP calls for, and, if not, why not.

Expressed more broadly, the main reasons for writing this book are threefold: first, because there is a shortage of research on the rural commuter belts of major cities, yet these are major loci of population change, economic growth and dynamic social change within city-regions, such that neither change within cities nor change in more traditional rural environments can be understood adequately without a solid appreciation of the rural commuter belt, its functioning and conflicting tendencies (e.g. Commission of the European Communities, 1999); second, because, as the ESDP shows, rural areas are central arenas within policy debates at the moment, with major readjustments in the cultural, economic and social condition of the countryside, and with policy debate rife over rural issues, yet with comparatively little in-depth analysis of the interaction (and mutual disturbance) of different sectors and segments within rural zones (e.g. Woods, 2005), which this book explores; and, third, because there is an ongoing need for comparative understanding on how similar processes within the European Union are manifest in both distinctive and shared ways across nations. In approaching these issues with an eye to the manner in which processes in different activity spheres or 'sectors' interact and disturb one another, the chapters of this book seek to move away from commonplace approaches to exploring cross-national differences in Europe, which commonly focus on single sectors or policy fields (e.g. Pfau-Effinger, 1994; Billaud *et al.*, 1997; Golland, 1998). This book is intended not simply to make statements about the nature of change in rural hinterlands but also to demonstrate appropriate viewpoints from which to interpret and understand the evolving diversity that characterizes rural Europe.

As for this chapter, its intentions are more modest, with three main issues targeted for attention. The first is to clarify what we mean by the city's hinterland, as compared with the cacophony of concepts like exurbs, peri-urban zones and urban fringe that overlap with this phrase. This commentary is not contemplative definitional musing, but seeks to highlight an absence of clarity over the geography of significant city-rural interactions. This is followed by an exploration of the policy context of hinterland areas, which pays particular attention to the ESDP, especially in establishing the principles policy-makers have specified as providing an acceptable basis on which the future advancement of city-regions is predicated (e.g. Faludi, 2003). This leads onto a closer examination of the themes that are explored in the research chapters that comprise the rest of the book.

**Rurality in the City's Hinterland**

The research project that led to this book started with the rather unpromising nomenclature of exploring the impact of urban centres on rural areas by way of investigating change in peri-urban areas. For those who are not aficionados of peri-urbanization, the incongruity of this quest might not be transparent, so it needs emphasizing that the concept 'peri-urban' neither suffers from universal usage, nor is there exactitude over its meaning. Of note here is the potential for confusion over the geographical extent of 'peri-urban' territory. Thus, embodied within Errington's (1994) claim that peri-urban zones are a forgotten European territory, is the heavy colouring of a definitional fix. It takes little time reading Errington's paper to grasp that the vision of peri-urban he presents is known more commonly as 'the rural-urban fringe'. The origin of this concept has been traced by Whitehand (1988) to the 1936 writings of the German geographer Herbert Louis.[2] An early US study of this 'fringe' defined it as '... that area of interpenetrating rural and urban land uses peripheral to the modern city' (Martin, 1953, p.iii). Similar phraseology is found in Bryant and associates' (1982, p.11) notion that the fringe is an area of transition between well-recognized urban land-uses and agricultural/ rural areas, or in Elson's (1987, p.19) decree that the urban fringe is '... the urban shadow, an area of fragmented and "intruded" farmland near the urban fence'. In

---

2  Audivac (1999) by contrast ascribes the idea to T.L. Smith in 1937, as the built-up area just outside the official limits of a city.

earlier work, Elson (1979) showed a further similarity with Martin, for he saw the urban fringe concept drawing attention to particular problems, like land-use transition and mixed land-uses. This insight is apparent in Martin's (1953) codifying the chaos of mixed land-uses that he associated with the fringe through descriptive attachments like 'marginal area' and 'twilight zone'. But if a complex interweaving of land-uses characterizes such areas, then this undoubtedly helps explain Errington's (1994) conclusion that fringe areas are little investigated in countries like Britain.

At one time urban fringe areas in the UK were zones of keen political and public interest, but the concern and antagonism that was generated by changes in these zones in the 1920s and 1930s led to strong opposition to their expansion (Lowerson, 1980; Jeans, 1990), which culminated in the passing of the London Green Belt Act in 1938 (Munton, 1983). With continued aggravation over the prospect of 'urban sprawl' and outward ribbon development, enactments of green belt protection became more common after the Second World War (Elson, 1986). Indeed, even though 1980s Conservative governments sought to water down other regulatory arms of the land-use planning system, public opposition restricted any desire to weaken green belt restraints (Bishop, 1998). Far from seeing a patchwork of land-uses in the rural-urban fringe, in Britain at least clinical lines of land-use distinction continued to be drawn between rural and urban territories. Those zones that had an uncertain land-use purpose in earlier decades found former confusion eradicated, either through infilling or by enforced land-use reversals (Hardy and Ward, 1984).

This does not mean that the immediate zones around the edge of built-up urban areas have not been of research interest, even if land-use planning keeps sharp lines of distinction between urban and rural (e.g. Ilbery, 1991). Yet the focus on this zone is rather restricted, owing to the limitations imposed by land-use planning. Hence, it has long been recognized that: 'Urban pressures are, for the most part, only of great significance to farmers' business practices on the immediate urban fringe' (Munton, 1974, p.216). Looking at activity spheres that extend beyond such limited effects, urban impacts on rural areas stretch across much more extensive territories. But do these constitute part of the 'urban fringe'? Capturing the indeterminacy with which this concept has been employed, Pryor (1968) holds that we need to distinguish between the urban-rural fringe, in which housing density is above average for the fringe as a whole, and the rural-urban fringe, where it is below the average. Readers might need to read the last sentence twice, for this writer at least finds the slippage from urban-rural to rural-urban an easy distinction to miss when 'fringe' is attached to the expression. Uncertainty over meaning is brought into further focus by some commentators interpreting the 'fringe' as an extensive hinterland zone. Thus, Lapping and Furuseth (1999, p.1) charge that, lying 40-50 miles (65-80km) from every major urban concentration, is a vast territory which can be recognized as the rural-urban fringe. Precisely this area is embodied in the term 'regional city' by Bryant and associates (1982, p.8), is captured in Herington's (1984) 'outer city' and is not far from designations of city-centred travel-to-work areas (Champion et al., 1987).

Moreover, in contrast to Errington's (1994) use of the phrase, in France such spacious geographical areas are known as 'peri-urban' zones. That said, as with the 'urban fringe', embodied within the French notion of a peri-urban zone is an understanding that there is a mingling of open country and built-up territory, with this mixed land-use zone occurring beyond suburban zones as you travel outwards from central cities (Cadene, 1990). However, as with their counterparts in other nations and continents, French researchers have been liberal in their interpretations of what peri-urban and peri-urbanization mean. One symbol of this is the varied terms that are used to explain the patterns, processes and consequences of peri-urbanization, with 'rurbanization' used to express processes of urban sprawl (Bauer and Roux, 1976), while other writers refer to 'exurbanization' or 'suburbanization' to capture the same processes and phenomena others call 'peri-urbanization'. In good part this reflects different interpretations of the processes and areas involved. For some, peri-urbanization is simply a form of urban sprawl (Nicot, 1995). More generally, peri-urbanization is linked to an increasing rural population in areas adjacent to cities and towns, alongside processes of social and functional recomposition within these areas. Such interpretations tend to view the urban-rural relationship as one-way, with rural areas considered by urbanites as sites of consumption (Bauer and Roux, 1976).

At least in so far as a variety of terms are used to signify such processes, the picture in Germany is similar to that in France. A key difference is that, like Britain, in Germany the concept 'peri-urban' has little currency. Instead, the terms dis-urbanization or exurbanization are more common. Similarly, in Spain the more common terminology is to refer to 'rururban' areas. What is meant by these notions is the growth of cities into their wider surroundings, alongside a relocation of city populations, and potentially spheres of administrative control, into the hinterland. In Germany, this process is conceptualized as a continuation of urbanization and suburbanization (e.g. Bär, 2003). From this perspective peri-urbanization is a form of suburbanization into new, outer rings (Heineberg, 2003). In Spain, the urban dispersion process has likewise occurred over zones with imprecise limits, where 'country' and 'city' ways of life are mixed, in territories that see the fastest and deepest population and morphological change of all urban spaces (Zárate, 1984, p.100). As with France and Germany, this expansion into non-urban and non-contiguous urban (or small town) space is associated with an intermingling of 'urban' and 'rural' land-uses as urban housing and economic activities penetrate into rural milieu (Chevalier, 1993; Heineberg, 2003).

When we look at the French, the German and the Spanish cases, we see elements of consistency in socio-cultural and landscape change in zones near urban edges (the urban fringe), as well as in districts that are more distant from the urban rim (i.e. more broadly across a city's rural hinterland). In the UK the intermingling of land-uses identified here is less common, as the land-use planning system keeps vigorous check on the incursion of 'urban land-uses' into 'the countryside'. Yet while they are to some extent 'invisible' as physical manifestations, the cultural, economic and social ties that bind rural and urban together in the UK are not only similar to those in France, Germany and Spain, but are arguably more potent. One good reason for this is the length of time (and geographical extent) over which city and hinterland have been bound together within the UK. Consider, for example, an observation made almost 100 years ago about South East England:

> In a manner all southeastern England is a single urban community; for steam and electricity are changing our geographical conceptions. A city in an economic sense is no longer an area covered continuously with streets and houses ... The metropolis in its largest meaning includes all the counties for whose inhabitants London is 'Town', whose men do habitual business there, whose women buy there, whose morning paper is printed there, whose standard of thought is determined there. East Anglia and the West of England possess a certain independence by virtue of their comparatively remote position, but, for various reasons, even they belong effectively to Metropolitan England. (Mackinder, 1907, p.258)

At the start of the twentieth century localized areas were clearly regarded as 'rural' within the regional landscape of South East England, with fears over the potential 'urban capture' of villages and hamlets (e.g. Bourne, 1912), yet as Mackinder astutely asserts the central features of the South East have long been driven by London. The existence of localized ruralities in urbanized regions points to the importance of scale in understanding the intensity and character of city-rural linkages. Smaller towns can have localized inter-relations with surrounding rural areas, irrespective of whether these towns are in rural regions or fall within the regional orbit of a major city. Of course, if the latter is the case, then the nature of urban-rural interactions can be expected to be more complex, much as the dispersal of population growth impulses from London has seen less direct movement from London to rural areas than shifts down the urban hierarchy in an outward city – town – rural flow (Warnes, 1991; Champion and Congdon, 1992).

One means of addressing such complexity is to recognize a hierarchy of (often urban-centred) 'functional regions', within which a distinction is drawn between the 'regional' scale and more 'localized' networks of exchange. For Britain, for example, travel-to-work areas identify regional scale journeys to work that have a shared focus on one city, with few zones left which merit the designation 'rural travel-to-work areas' because their commuter flows are dispersed across a series of small towns and villages (Coombes *et al.*, 1982; Atkins *et al.*, 1996). Yet, if we take Connell's (1974) suggestion, that an urban commuter belt might be comprised of areas from which at least 20 per cent of the working population travel to work, then we can anticipate more localized travel-to-work areas, which are often centred on one or more small towns (Table 1.1). Add into this equation what urban 'catchment' areas might look like for other economic or social activities, and the ease of classifying coherent 'city-regions' stands on shakier ground (Coombes, 2000). Hence, when writing on what constitutes 'exurban', Bell (1992, p.67) argues that: '... a certain indeterminacy is perhaps central to their definition. What I mean by exurbs is that region where plenty of city money is to be had, but where pastures, fields, woods, or other forms of rural enterprise clearly dominate the landscape. Exurbs, then, are areas where people likely argue from time to time if this is really still the country'. That the issue of whether village and countryside are 'really rural' characterizes discussions in England as much as in other parts of

Europe, and has done so for some time (e.g. Anderson and Anderson, 1965; Spindler, 1973), bears testimony to the dominant position of identity and linkage over physical form in the determination of the city's hinterland.

**Table 1.1   Mean Average Values for Variables Used to Classify Rural NUTS5 Areas into Journey-to-Work Zones in Eastern England**

| Rural ward type | Distance To nearest city (km) | % 'local' workers | % working in main city | % working in secondary cities | Number of rural NUTS5 areas |
|---|---|---|---|---|---|
| Single-city centred commuter belts | 9.2 | 22.5 | 43.8 | 6.0 | 125 |
| Multi-city centred commuter belts | 11.0 | 25.1 | 20.4 | 19.8 | 106 |
| Dispersed commuting areas | 16.0 | 35.1 | 18.5 | 6.2 | 143 |
| Rural zones near mid-sized towns | 26.5 | 23.2 | 8.9 | 4.3 | 109 |
| 'Self-contained' rural areas | 33.3 | 41.6 | 5.0 | 2.3 | 113 |

Note:   These computations are based on 1991 Census data. 'Local workers' have paid employment in the NUTS5 area in which they live. 'Main' city refers to the city receiving the largest number of commuters from a rural NUTS5 area. 'Cities' are built-up areas with at least 50 000 inhabitants. 'Rural' wards are those with a population density of less than 400 per square kilometre. Eastern England is comprised of Bedfordshire, Cambridgeshire, Essex, Hertfordshire, Norfolk and Suffolk.

Critically, then, in this book we are not seeking to provide exactitude over what is the city's hinterland. Terms like exurbs, peri-urban, urban fringe and the like have a value in drawing attention to particular processes or causal linkages, but their usage is tempered by national custom, personal preference and the content of the argument being made. Providing a common method for demarcating European city hinterlands is far from an easy task, even if there were agreement on what kind of linkages are most fundamental to city-hinterland interactions (e.g. Cheshire et al., 1996). Of course, one element of these interactions is their potentially uneven importance across nations; the exploration of which is one of the objectives of this book. In order to explore whether dissimilar dimensions of rural-urban linkages make significant impressions on rural and urban realms, it is important to have an open-mind about the geographical spread for significant city-hinterland linkages. This does not mean this book approaches rural-urban interactions in the same manner as it views globalization processes. For sure, the Wall Street Crash of 1929 had a severe impact on rural Europe, just as it did on cities (Stevenson and Cook, 1977). This does not mean that the financial networks linking rural zones to events in New York constitute urban-rural linkages in the same manner as a city acts as a service centre for it primary hinterland. The nature of interactions between city and hinterland are more diverse, while retaining significant depth, than those that bind rural areas in one part of the world to markets, political decisions and specific places continents away (e.g. Sanderson, 1986). The city-regions that are the focus of attention in this book are of a kind that binds cultural, economic, political and social fortunes together within central cities and their primary hinterland territories.

## City-Rural Linkages and City-Region Fortunes

'Traditional' models of the diffusion of growth impulses envisage cities as centres of dynamism, with rural areas benefiting from urban spread effects (e.g. Berry, 1970). The simplicity of this model is now recognized as a serious shortcoming. For one, urban growth in the context of impoverished rural regions is quite a different process from that experienced alongside vibrant rural economies. Particularly in southern Europe, the apparent strength of some cities in terms of demographic growth results more from a lack of rural options than urban dynamism. Far from spreading growth benefits into the countryside, cities can be weighed down by rural impoverishment, so their problems are intensified by in-migration from the surrounding countryside (Cheshire and Hay, 1989). Yet city problems can also emanate from out-migration to surrounding communities. Much as in the USA (Berry, 1980), a 'suburbanization' process has been occurring in western European cities that has been accompanied by many city cores losing population, with contiguous suburban 'rings' seeing a slow down in population growth (Dematteis, 1998). Fundamental to this new demographic and economic geography of city-regions is a vision of urban influence extending outwards. Here the analogy of a bulldozer has been used by some (Audivac, 1999), although Hart (1991) prefers the imagery of a bow-wave, wherein the pressures from the city on surrounding rural areas are akin to a ship pushing waves forward as it moves through the water.

This pattern of changing urban fortunes has been captured by some in rather unidimensional models of urban growth (e.g. Berg *et al.*, 1982). Although compressing 'reality' rather hard, such models focus attention on transitions over time, as well as highlighting key dimensions of temporal change. Yet models of this kind are most advances in the USA, where processes of unfettered metropolitan decentralization started much earlier than in Europe. As Leinberger (1996) explained the US situation, metropolitan expansion occurred in four (or possibly five) stages. The first generation of growth had the central city as the focus for expansion. By the 1960s and 1970s, a second growth generation had begun, wherein expansion occurred primarily 3-10 kilometres from downtown, largely around freeways. This outward flow strengthened over time, such that the third generation was characterized by the emergence of new suburban centres adjacent to large tracts of upper-middle class and even higher income housing, with manufacturing growth located even farther from the central city. In the fourth generation, which many commentators see today, the growth foci are farther from downtown (some 6-20km beyond third generation centres), with corridors of office and manufacturing activity in low-density semi-rural areas (the prospect of a fifth generation is seen to have started with major warehouse developments even farther from the city).³ This is not to say that all cities are experiencing this process, for the relationship between the central city and its 'suburban' development is varied (Stanback, 1991); with the emergence of new 'suburban' centres that compete with central cities being affected by city size and growth dynamics. Nevertheless, where (discontinuous) 'suburban' centres have gained a significant hold, growth pressures on rural areas and small towns have intensified, with commuting and city-countryside links becoming more dispersed. Lifestyles have shifted in focus from the central city to individualized linkages with a variety of competing nodes in contiguous and non-contiguous realms of the city's hinterland (Fishman, 1987; Audivac, 1999). In other words, 'suburban' centres that appear to be subordinate to a central city might exert considerable independent influence on rural areas and small towns once inter-linkages are explored, even to the extent of diminishing the central city's influence (Garreau, 1991).

Yet the pace of change in hinterland areas is not uniform as it is significantly influenced by the competitiveness of the city-region as a whole. Hence, in the UK:

> Whereas the smaller settlements within large TTWAs [travel-to-work areas] may have seen the arrival within their boundaries of jobs decentralizing from nearby cities, this process of job movement seems to represent little change in the broader measure of accessible jobs for smaller town residents because it also involves lost jobs in the accessible larger settlements nearby. (Coombes and Raybould, 2004, p.218)

Hardly surprisingly, faced with such dynamic rural-urban inter-relationships, public authorities in a number of European countries have come to recognize the need for a strategy for the development of city-regions as a whole, rather than favouring development in cities and assuming this will bring benefits for hinterland zones (e.g. Adam, 2003). Significantly, this drive for increasing cooperation to promote city-region competitiveness has drawn noteworthy sustenance from European-level policy initiatives that have been both praised for their assertion of the need for more binding interaction between rural and urban zones within city-regions, at the same time as they have been condemned for heightening inter-regional inequalities under the guise of equalizing opportunities across European space.

*The European Policy Context*

Under the European Commission's (1999) European Spatial Development Perspective (ESDP), we are presented with a vision for the future that envisages more economic development outside the core European axis that runs from South East England through northern France and southern Germany into northern Italy (viz. the European 'Pentagon'). The emphasis in this vision is on polycentric development, which for city-region competitiveness refers to enhancing non-core city-regions using sustainable development solutions. Thus, the ESDP vision for

---

3  See Chapter Three for a review of similar ideas on the outward movement of development rings around cities in Germany.

city-regions is for them to be competitive internationally (Commission of the European Communities, 1999, p.22). Critics of this aim are keen to point out that the ESDP is little more than a list of policy options, yet, as Faludi (2003) indicates, it has become a standard reference against which spatial policy in the European Union is assessed (Jensen and Richardson, 2004). Yet critiques from the southern and northern extremities of the continent have cautioned over assuming that the ESDP presents a blueprint for development in peripheral city-regions (e.g. Richardson, 2000; Hadjimichalis, 2003). Critiques hold that the ESDP fuses 'rural' with 'peripheral', thereby confusing what is possible in the rural zones of city-regions in the European 'Pentagon' with what is achievable in peripheral, less densely settlement areas. This book is not the place to go into such criticisms in detail, but it is pertinent to note that commentators have identified a new logic to the location of business services. It is not simply that these growth-inducing services tend to be concentrated in major cities (e.g. Bennett *et al.*, 1999), but also that operations outside major clusters are less dynamic and less well integrated into global markets (e.g. Keeble and Nachum, 2002). Hence, a localization of activities is taking place, tending to strengthen the largest city-regions, with inter-(city) regional competition a powerful factor in the spatial division of labour and the reconfiguration of the urban hierarchy (Cheshire and Gordon, 1995a, 1995b). In this context the ESDP has so far not offered policy visions that offer counterweights to dominant tendencies within an increasing globalized political economy.

Yet, at a more localized scale, the ESDP does contain policy emphases with more potential for effective, practical initiatives. At this localized level, the aims of the ESDP come across as more feasible, as well as recognizing distinctions between 'core' and 'peripheral' city-regions:

> The future of many rural areas is becoming increasingly related to the development of urban settlements in rural areas. Towns and cities in rural regions are an integral component in rural development. It is essential to ensure that town and country can formulate and successfully implement regional development concepts in partnership based collaboration. However, the rural-urban relationship in densely populated regions differs from that in sparsely populated regions. In densely populated regions, the areas with rural characteristics are under substantial urbanization pressure, with all the side effects of increased density, including the negative ones. These include pollution of soil and water, fragmentation of open areas and the loss of rural character. Some traditional rural functions such as extensive agriculture, forestry, nature conservation and development, for example, are highly dependent on a high degree of continuous open countryside. A key function of spatial development is, therefore, to achieve a better balance between urban development and protection of the open countryside. Urban and rural areas are closely interconnected, especially in densely developed regions. Rural areas benefit from the cultural activities of cities, while the cities benefit from the leisure and recreation value of rural areas. Town and country are, therefore, partners rather than competitors. (Commission of the European Communities, 1999, p.66)

The image that sustainable city-regions require '... partnership between towns and cities of every size and their surrounding countryside' (Commission of the European Communities, 1999, p.25) is certainly not restricted to the ESDP. In the

UK, for example, Ewen Cameron, the former Chair of the Countryside Agency, has said that the national government's Urban and Rural White Papers must go hand in hand (Caffyn and Dahlström, 2001). Recognizing that rural and urban cannot be treated as distinctive, that they are integrated, in a mutually dependent and reinforcing interchange, which ensures that urban problems are also rural problems, and vice versa, is only part of the story. As important as urban-rural relationships are inter-sectoral links, given the growing '... realization that predominantly sectoral policies (agriculture, environment, transport, housing and tourism) are no longer able to deliver policies appropriate to the changing needs of the countryside' (Hadjimichalis, 2003, p.108), nor indeed for cities.

In effect the European Commission is promoting joined-up thinking in both geographical and sectoral terms, acknowledging that there is mutuality in the capacity for friction in either of these spheres to disturb progress toward enhancing the competitiveness of city-regions and improving the quality of life of local residents. In this context, a primary aim in the research that underpins the chapters of this book is to investigate how far trends within city-regions are creating tensions that work against the realization of quality of life increments. Taking its cue from dimensions articulated in the ESDP, the central issues investigated herein are how the quality of life of citizens, especially in hinterland zones, are impacted on by:

- The physical expansion of urban centres
- Changes in social and functional heterogeneity (especially as related to social exclusion)
- Improvements in transport accessibility
- The conservation and development of natural and cultural heritage.

Adding clarity to what the European Commission envisages as the encompassing requirements of successful sustainable development, the ESDP makes clear that meeting its objectives at a city-region level requires:

- Pursuit of 'compact city' settlement forms in order to control urban expansion
- The reconstruction of derelict areas, appropriate access to basic services and facilities for all citizens, and more open (green) spaces
- An integrated approach 'with closed cycles' for natural resource usage, energy and waste disposal
- The location of land-uses and integrated transport planning in order to reduce car usage (Commission of the European Communities, 1999, pp.22-23).

In order to achieve such objectives, collaboration and cooperation between regulatory (and other) agencies are important, with such exchanges requiring:

- Equality and independence between partners
- Voluntary participation
- Consideration of different administrative conditions
- Common responsibility and benefits.

The chapters that follow are designed to investigate how far these principles of 'good practice' are integral to processes of change within city hinterlands and, if they are, how far these can be viewed as desirable goals?

This latter is a point that needs emphasizing, for while the ESDP (and other governmental policy aims) place stress on the achievement of goals like those listed above, experience in some countries has shown that some of these goals are of questionable merit. In the case of the Netherlands, for example, the Fifth National Policy Document has questioned the previous planning emphasis on compact cities (Burg and Dieleman, 2004). Today, the growing size of daily urban systems, the inadequacies of public transport provision, and the manner in which global competition can favour small over large places, are acknowledged as weaknesses in compact city formulations; with citizen preferences for rural living creating a 'democratic deficit' that works against public support for such policies (Dam *et al.*, 2002). With acceptance that society is now more network-based than area-based in its relationships (Castells, 2000), this points to the possible need for more flexibility in policy principles, to strategic vision without a specific blueprint. How far this view will hold in other nations is an issue explored herein.

Of course, the characterization of change in city hinterlands is likely to differ across the nations explored in this book. This is a consistent message from research on the European space economy, such as that by Pompili (1994) and Rodríguez-Pose (1994), which show a nation-specific impetus to development patterns. Already this chapter has noted that the manner and extent of state intervention in the regulation of land-use have marked bearing on the form and evolution of urban structures, as well as their relationships with surrounding countryside areas. Two extremes on this spectrum are the United States and the former communist states of eastern Europe. In the USA government regulation of urban growth has been ineffective, in good part because the planning system has long been a pawn in national and local political economies (Gottdiener, 1977; Heiman, 1988). As a result city-regions have easily drifted toward a multi-focal structure, with urban expansion sprawling into the surrounding countryside, as wave upon wave of outward expansion imposes chaotic, multi-centred growth impulses. In the USA central cities are increasingly divorced from surrounding suburban and rural areas (Garreau, 1991; Audivac, 1999). Communist eastern Europe sharply contrasted with the USA, with government regulation of work and home locations binding their proximity, until the fall of the Berlin Wall generated profound space economy change (e.g. Wild and Jones, 1994; Vogeler, 1996). But while such extremes point to the power of government regulation over urban change, other 'extremes' warn us that too much weight should not be ascribed to government direction. This is very evident in South Africa, where the abolition of apartheid has had a muted effect on urban segregation (Christopher, 1994; Roberts, 1994). Exploring city-rural interchanges from a shared concern about the merits of ESDP priorities, the chapters that follow offer genuine possibilities for evaluating cross-national differences and similarities, which is the issue that is turned to in the last chapter.

## Themes Explored on City Hinterlands

This book is based on a coordinated programme of research into trends, conflicts and outcomes associated with change in the rural hinterlands of cities in four European nations (England, France, Germany and Spain). Taking ESDP principles as its starting point, the aim is to question whether rural areas and small towns are integrated effectively into quality of life improvements as a result of stronger inter-dependencies with cities. The book analyses this hope, focusing on the critical arenas of employment change, housing and service provision. In doing so, it investigates how change in these fields impact on the quality of life and physical environment of rural hinterlands. The book shows that often assumed distinctions between European nations are less potent when put under the microscope, with many shared problems for hinterland zones. Yet diversity exists, often for reasons that are not obvious, although with messages that have important implications for conceptualizing and theorizing change within city-regions. Lessons can be derived from this, which mean that tendencies in one country can embody powerful messages for policy-makers in other nations who are seeking 'successful' development outcomes.

The chapters that follow begin with an examination of rural diversity within the European Union, paying attention to the identification and characterization of city commuter belts. This chapter notes difficulties in securing a pan-European picture of areas that come under the direct influence of city centres, for there are conceptual and methodological difficulties in mapping and describing such territories in a uniform manner. Yet broad trends can be specified and diversity within these trends identified. The focus of this second chapter is on the nature of that diversity, so this chapter provides a contextualization for the nation-specific case studies that follow.

The next four chapters focus on specific city-regions within each of England, France, Germany and Spain. The city-regions explored have been chosen to draw out different aspects of city-hinterland linkages, with 'intervening' forces injected diversity in change forces in some areas (e.g. tourism expansion in parts of rural hinterlands and transnational movements for work and second home occupancy). Each of these chapters follows the same pattern in exploring key changes within hinterland territories, with particular regard to trends and associated conflicts over employment, housing and services. These chapters identify dominant change forces, and assess their consequences for the achievement of ESDP goals of enhancing sustainable development, mitigating social exclusion and promoting more efficacious land-use change. Coherence, and difference, is offered across these chapters, because they combine treatment of the same issues with an analytical focus that is embedded in environments in which key influences on hinterland change exhibit dissimilar nuances and emphases (e.g. some areas are seeing rapid economic growth, others are more sluggish, some have tight land-use planning controls, others have lax ones or ones in which other priorities conflict with land-use planning ideals).

In terms of the main case study sites for the in-depth investigations undertaken, their selection laid most stress on intermediate sized city-regions, away from national economic cores, although only one of the city-regions investigated is well away from the continental economic centre. With (approximate) populations for the (administrative areas that comprise the) central cities in these city-regions, the main study areas are:

- France     Annecy Urban Area 142 000 (secondary area Valence)
- Germany    Munich 1 200 000
- Spain      Granada 240 000 (secondary area El Ejido)
- UK         Norwich 175 000 (secondary areas Bedford and Cambridge).

An obvious difference in these study areas is between Munich, which is a major centre in the German political economy, with a particularly dynamic economy that pushes the Bavarian economy forward, and Norwich, which is relatively slow growing in a UK context, and presents itself as a middle-sized provincial centre of regional importance but without a dynamic economy. These two centres represent the extremes in the study areas in terms of economic and demographic change (although Valence is close to the Norwich position), at least in terms of positions in national urban hierarchies. Yet all these city-regions are seeing increasing pressure for new housing and, in varying degrees, for new employment nodes in hinterland zones. In Annecy and Granada, for example, added to centripetal forces within city-regions themselves are infusions of demand from tourism and trans-border movements (e.g. France offering cheaper housing than nearby Switzerland). Hence, even though Granada is in a region with a traditionally sluggish economy (Salmon, 1992), there are strong growth tendencies within the city's hinterland. What such pressures mean for social, housing, environmental, employment and cultural change in each of the city-regions is explored in turn for the German (Chapter Three), French (Chapter Four), Spanish (Chapter Five) and English (Chapter Six) case studies. These chapters explore how tendencies in city hinterlands ease or retard the achievement of ESDP and other European Union goals, as with increased social exclusion from in-migration, greater environmental problems resulting from outmigration and farm abandonment, an imbalance in employment prospects owing to economic change, difficulties of accessing services and so on.

The final chapter offers a comparative perspective on the national case studies. This chapter is designed to explore the degree to which trends in each nation, along with the conflicts and tensions exposed, are expressed in similar or divergent ways in different national contexts. In the context of European Commission desires to enhance city-region competitiveness, especially away from the continental core, this chapter investigates how far cross-national similarities and differences can be traced to peculiar combinations of cultural, economic and social circumstances, or whether such trends are nation-specific. Given that, as Faludi (2003) identifies, ESDP principles are broadly accepted, even if unevenly implemented, across the European Union, the diversity of contexts in which the case studies for this book are located provides an appropriate framework for exploring whether the development of harmonious partnerships between rural and urban are feasible in the immediate future.

# References

Adam, B. (2003) Spatial policies for metropolitan regions – identity, participation and integration, *European Planning Studies*, 11, 739-747

Anderson, R.T. and Anderson, B.G. (1965) *Bus Stop for Paris: The Transformation of a French Village*, Doubleday, Garden City, New York

Atkins, D.J., Champion, A.G., Coombes, M.G., Dorling, D.F.L. and Woodward, R. (1996) *Urban Trends in England: Latest Evidence from the 1991 Census*, HMSO, London

Audivac, I. (1999) Unsettled views about the fringe: rural-urban or urban-rural frontiers?, in O.J. Furuseth and M.B. Lapping (eds.) *Contested Countryside: The Rural Urban Fringe in North America*, Ashgate, Aldershot, 7-32

Bär, J. (2003) Entwicklung von Urbanisierung, at http://www.berlin-institut.org/pdfs/Baehr_Urba-nisierung_Entwicklung.pdf

Bauer, G. and Roux, J. (1976) *La rurbanisation ou la ville éparpillée*, Seuil, Paris

Bell, M.M. (1992) The fruit of difference: the rural-urban continuum as a system of identity, *Rural Sociology*, 57, 65-82

Bennett, R.J., Graham, D.J. and Bratton, W.J.A. (1999) The location and concentration of businesses in Britain, *Transactions of the Institute of British Geographers*, 24, 393-420

Berg, L. van der, Drewett, R., Klaasen, L.H., Rossi, A. and Vijverberg, C.H.T. (1982) *Urban Europe: A Study of Growth and Decline*, Pergamon, Oxford

Berry, B.J.L. (1970) Commuting patterns: labour market participation and regional potential, *Growth and Change*, 1(4), 3-10

Berry, B.J.L. (1980) Inner city futures: an American dilemma revisited, *Transactions of the Institute of British Geographers*, 5, 1-28

Billaud, J-P., Bruckmeier, K., Patricio, T. and Pinton, F. (1997) Social construction of the rural environment, in H.J. de Haan, B. Kasimis and M.R. Redclift (eds.) *Sustainable Rural Development*, Ashgate, Aldershot, 9-34

Bishop, K.D. (1998) Countryside conservation and the New Right, in P. Allmendiger and H. Thomas (eds.) *Urban Planning and the New Right*, Routledge, London, 186-210

Bourne, G. [pseudonym for Sturt, G.] (1912) *Change in the Village*, 1955 edition, Duckworth, London

Bryant, C.R., Russwurm, L.H. and McLellan, A.G. (1982) *The City's Countryside: Land and its Management in the Rural-Urban Fringe*, Longman, London

Burg, A.J. van der and Dieleman, F.M. (2004) Dutch urbanization policies: from 'compact city' to 'urban network', *Tijdschrift voor Economische en Sociale Geografie*, 95, 108-116

Cabus, P. and Vanhaverbeke, W. (2003) The economics of rural areas in the proximity of urban networks: evidence from Flanders, *Tijdschrift voor Economische en Sociale Geografie*, 94, 230-245

Cadene, P. (1990) L'usage des espaces périurbains, une géographie régionale des conflits, *Etudes Rurales*, 118/119, 225-267

Caffyn, A. and Dahlström, M. (2001) *Connecting Town and Country: A Study of Joint Working and Partnerships Between Urban and Rural Areas in England*, University of Birmingham Centre for Urban and Regional Studies West Midlands Research Report 18, Birmingham

Castells, M. (2000) *The Rise of the Network Society*, second edition, Blackwell, Oxford

Champion, A.G. and Congdon, P. (1992) Migration trends for the South: the emergence of a Greater South East?, in J. Stillwell, P.H. Rees and P. Boden (eds.) *Migration Processes and Patterns: Volume Two - Population Redistribution in the United Kingdom*, Belhaven, London, 178-201

Champion, A.G., Green, A.E., Owen, D.W., Ellin, D.J. and Coombes, M.G. (1987) *Changing Places: Britain's Demographic, Economic and Social Complexion*, Edward Arnold, London

Cheshire, P.C. and Gordon, I.R. (1995a) European integration: the logic of territorial competition and Europe's urban system, in J. Brotchie, M. Batty, P.G. Hall and P. Newton (eds.) *Cities in Competition: Productive and Sustainable Cities for the 21st Century*, Longman Australia, Melbourne, 108-126

Cheshire, P.C. and Gordon, I.R. (eds., 1995b) *Territorial Competition in an Integrating Europe*, Avebury, Aldershot

Cheshire, P.C. and Hay, D.G. (1989) *Urban Problems in Western Europe*, Unwin Hyman, London

Cheshire, P.C., Furtado, A. and Magrini, S. (1996) Quantitative comparisons of European regions and cities, in L. Hantrais and S. Mangen (eds.) *Cross-National Research Methods in the Social Sciences*, Pinter, London, 39-50

Chevalier, M. (1993) Neo-rural phenomena, in *Two decades of l'Espace Géographique: Special Issue in English*, Reclus, Paris, 175-191 (published in French in *l'Espace Géographique* 10 (1981) 33-47)

Christopher, A.J. (1994) Segregation levels in the late-Apartheid city 1985-1991, *Tijdschrift voor Economische en Sociale Geografie*, 85, 15-24

Commission of the European Communities (1999) *ESDP – European Spatial Development Perspective: Towards Balanced and Sustainable Development of the Territory of the European Union*, Office for Official Publications of the European Communities, Luxembourg

Connell, J. (1974) The metropolitan village: spatial and social processes in discontinuous suburbs, in J.H. Johnson (ed.) *Suburban Growth*, Wiley, London, 77-100

Coombes, M.G. (2000) Defining locality boundaries with synthetic data, *Environment and Planning*, A32, 1499-1518

Coombes, M.G. and Raybould, S. (2004) Finding work in 2001: urban-rural contrasts across England in employment rates and local jobs availability, *Area*, 36, 202-222

Coombes, M.G., Dixon, J.S., Goddard, J.B., Openshaw, S. and Taylor, P.J. (1982) Functional regions for the population census of Great Britain, in D.T. Herbert and R.J. Johnston (eds.) *Geography and the Urban Environment: Volume Five*, Wiley, Chichester, 63-112

Dam, F. van, Heins, S. and Elbersen, B.S. (2002) Lay discourses of the rural and stated and revealed preferences for rural living: some evidence of the existence of a rural idyll in the Netherlands, *Journal of Rural Studies*, 18, 461-476

Dematteis, G. (1998) Suburbanización y periurbanización. Ciudades anglosajonas y ciudades latinas, in F.J. Monclús (ed..) *La ciudad dispersa. Suburbanización y nuevas periferias*. Centre de Cultura Contemporània de Barcelona, Barcelona (also at http://www.ub.es/geocrit/aracne.htm)

Dunford, M. and Perrons, D. (1994) Regional identity, regimes of accumulation and economic development in contemporary Europe, *Transactions of the Institute of British Geographers*, 19, 163-182

Elson, M.J. (1979) *The Leisure Use of Green Belts and Urban Fringes*, Social Science Research Council and The Sports Council, London

Elson, M.J. (1986) *Green Belts: Conflict Mediation in the Urban Fringe*, Heinemann, London

Elson, M.J. (1987) The urban fringe – will less farming mean more leisure?, *The Planner*, October, 19-22

Errington, A.J. (1994) The peri-urban fringe: Europe's forgotten rural areas, *Journal of Rural Studies*, 10, 367-375

Faludi, A. (2003) Unfinished business: European spatial planning in the 2000s, *Town Planning Review*, 74, 121-140

Faludi, A. and Waterhout, B. (2002) *The Making of the European Spatial Development Perspective: No Masterplan*, Routledge, London

Fishman, R. (1987) *Bourgeois Utopias: The Rise and Fall of Suburbia*, Basic, New York

Garreau, J. (1991) *Edge City: Life on the New Frontier*, Anchor, New York

Golland, A. (1998) *Systems of Housing Supply and Housing Production in Europe: A Comparison of the United Kingdom, the Netherlands and Germany*, Ashgate, Aldershot

Gottdiener, M. (1977) *Planned Sprawl: Private and Public Interests in Suburbia*, Sage, Beverly Hills, California

Hadjimichalis, C. (1994) The fringes of Europe and EU integration: a view from the South, *European Urban and Regional Studies*, 1, 19-30

Hadjimichalis, C. (2003) Imagining rurality in the New Europe and dilemmas for spatial policy, *European Planning Studies*, 11, 103-113

Hardy, D. and Ward, C. (1984) *Arcadia for All: The Legacy of a Makeshift Landscape*, Mansell, London

Hart, J.F. (1991) The perimetroplitan bow wave, *Geographical Review*, 81, 35-51

Heiman, M.K. (1988) *The Quiet Revolution: Power, Planning and Profits in New York State*, Praeger, New York

Heineberg, H. (2003) Grundriß Allgemeine Geographie: Stadtgeographie, at http://www.millennium.fran-ken.de/fsi/html/skripte/HeinebergStadtgeo.pdf

Herington, J. (1984) *The Outer City*, Harper and Row, London

Herrschel, T. and Newman, P. (2002) *Governance of Europe's City Regions: Planning, Policy and Politics*, Routledge, London

Ilbery, B.W. (1991) Farm diversification as an adjustment strategy on the urban fringe of the West Midlands, *Journal of Rural Studies*, 7, 207-218

Jeans, D.N. (1990) Planning and the myth of the English countryside, in the interwar period, *Rural History*, 1, 249-264

Jensen, O.B. and Richardson, T. (2004) *Making European Space: Mobility, Power and Territorial Identity*, Routledge, London

Keeble, D.E. and Nachum, L. (2002) Why do business service firms cluster? Small consultancies, clustering and decentralization in London and southern England, *Transactions of the Institute of British Geographers*, 27, 67-90

Keeble, D.E., Lawson, C., Moore, B. and Wilkinson, F. (1999) Collective learning processes, networking and 'institutional thickness' in the Cambridge region, *Regional Studies*, 33, 319-332

Keeble, D.E., Tyler, P., Broom, G. and Lewis, J. (1992) *Business Success in the Countryside*, HMSO, London

Lapping, M.B. and Furuseth, O.J. (1999) Introduction and overview, in O.J. Furuseth and M.B. Lapping (eds.) *Contested Countryside: The Rural Urban Fringe in North America*, Ashgate, Aldershot, 1-5

Leinberger, C.B. (1996) Metropolitan development trends of the late 1990s: social and environmental implications, in H.L. Diamond and P.F. Noonan (eds.) *Land Use in America*, Island Press, Washington DC, 203-222

Lowerson, J. (1980) Battles for the countryside, in F. Gloversmith (ed.) *Class, Culture and Social Change: A New View of the 1930s*, Harvester, Brighton, 258-280

Mackinder, H.J. (1907) *Britain and the British Seas*, second edition, Clarendon, Oxford (1930 reprint)

Martin, W.T. (1953) *The Rural-Urban Fringe: A Study of Adjustment to Residence Location*, University of Oregon Press, Eugene

Morgan, K. (1997) The learning region: institutions, innovation and regional renewal, *Regional Studies*, 31, 491-503
Munton, R.J.C. (1974) Farming on the urban fringe, in J.H. Johnson (ed.) *Suburban Growth*, Wiley, London, 210-223
Munton, R.J.C. (1983) *London's Green Belt: Containment in Practice*, Allen and Unwin, London
Nicot, B-H. (1995) *La périurbanisation dans les zones de peuplement industriel et urbain*, Université de Paris XII SIRIUS Papier 95-36, Créteil
Pfau-Effinger, B. (1994) The gender contract and part-time paid work by women - Finland and Germany compared, *Environment and Planning*, A26, 1355-1376
Pompili, T. (1994) Structure and performance of less developed regions in the EC, *Regional Studies*, 28, 679-693
Pryor, R.J. (1968) Defining the rural-urban fringe, *Social Forces*, 47, 202-215
Richardson, T. (2000) Discourses of rurality in EU spatial policy: the European Spatial Development Perspective, *Sociologia Ruralis*, 40, 53-71
Roberts, M. (1994) The ending of apartheid: shifting inequalities in South Africa, *Geography*, 79, 53-64
Rodríguez-Pose, A. (1994) Socio-economic restructuring and regional change: rethinking growth in the European Community, *Economic Geography*, 70, 325-343
Salmon, K.G. (1992) *Andalucía: An Emerging Regional Economy in Europe*, Junta de Andalucía Consejería de Economia y Hacienda, Sevilla
Sanderson, S.E. (1986) The emergence of the 'world steer', in F.L. Tullis and W.L. Hollist (eds.) *Food, the State and the International Political Economy*, University of Nebraska Press, Lincoln, 123-148
Spindler, G.D. (1973) *Burgbach: Urbanization and Identity in a German Village*, Holt, Rinehart and Winston, New York
Stanback, T.M. (1991) *The New Suburbanization: Challenge to the Central City*, Westview, Boulder
Stevenson, J. and Cook, C. (1977) *The Slump: Society and Politics During the Depression*, Jonathan Cape, London
UK Royal Commission on Local Government in England (1969) *Report*, (Redcliffe-Maud Report) HMSO, London
Vogeler, I. (1996) State hegemony in transforming the rural landscapes of eastern Germany: 1945-1994, *Annals of the Association of American Geographers*, 86, 432-458
Warnes, A.M. (1991) London's population trends: metropolitan area or megalopolis?, in K. Hoggart and D.R. Green (eds.) *London: A New Metropolitan Geography*, Edward Arnold, London, 156-175
Whitehand, J.W.R. (1988) Urban fringe belts: development of an idea, *Planning Perspectives*, 3, 47-58
Wild, T. and Jones, P.N. (1994) Spatial impacts of German unification, *Geographical Journal*, 160, 1-16
Woods, M. (2005) *Contesting Rurality: Politics in the British Countryside*, Ashgate, Aldershot
Zárate, A. (1984) *El mosaico urbano: organización interna y vida en las ciudades*, Editorial Cincel, Madrid

## Chapter 2

# Diversity in the Rural Hinterlands of European Cities

Vincent Briquel and Jean-Jacques Collicard

**Introduction**

Urban sprawl is blurring borders between cities and neighbouring rural areas, which are losing their traditional appearance with the establishment of new infrastructure and activities, and the arrival of new inhabitants (Steinberg, 1991; Prost, 1993). Consequently, the influence exerted today by urban centres on rural areas, whether densely populated or not, is increasing, and is the sign of a growing interdependence between these areas. This trend may be observed throughout the European Union, including its new Member States. Rural areas are being gradually integrated in environments structured by cities (Jaillet and Jalabert, 1982; Berger *et al.*, 1980) and this integration is resulting in new modes of development for these areas.

European policy-makers have recognized the new potentialities and challenges this creates. As early as 1994, Ministers responsible for spatial planning agreed on policy guidelines for the spatial development of the EU, one of which is enhancing urban-rural partnerships to overcome outdated dualisms between city and countryside (Faludi and Waterhout, 2002). Thus, the European Spatial Development Perspective (ESDP) stresses the future benefits of greater rural-urban integration in terms of balanced development for the European territory (Commission of the European Communities, 1999, §65). From an ESDP viewpoint, competitive towns and cities will base their future prospects in part on surrounding rural areas (§81), with town and countryside required to share an integrated approach since they form a region and are mutually responsible for its further development (§101). Thus, the ESDP identifies key spatial planning policy options, as '... promotion of co-operation between towns and countryside aiming at strengthening functional regions' and '... integrating the countryside surrounding large cities in spatial development strategies for urban regions, aiming at more efficient land-use planning, paying special attention to the quality of life in the urban surroundings' (Commission of the European Communities, 1999, §106). As part of city-region development, town and city need to contribute to general quality of life conditions, as expressed, for example, by controlling their outward physical expansion and promoting environmentally-sound management processes for urban eco-systems (§81), both of which the ESDP views as pre-conditions for the sustainability of urban-centred regions.

The ESDP has certainly captured forces at work in rural areas near cities, which have been integrated into so-called urban regions comprised of central cities and their areas of primary influence (viz. their hinterlands). The network of linkages so resulting form a structure that is dense enough for the whole territory to have urban economic and social characteristics, as well as functionally constituting a unitary space (Precedo, 1988; Pumain and Saint-Julien, 1996). This does not mean that the processes at play are sufficiently understood. Thus, the goal of the ESPON Project, which was launched following the study programme on European spatial planning that accompanied the ESDP (Study Programme on European Spatial Planning, 2000), is to observe and analyze trends in territorial and regional development in Europe, notably via thematic studies of territorial effects of major forces in spatial development (Coll *et al*., 2000). Picking up a strong theme in the ESDP, the ESPON (2004) study on the potential for polycentric development focuses on the concept of functional urban areas (FUA) or regions (Antikainen and Vartiainen, 2002; OECD, 2002), which is used to analyze regional development at a spatial level that distinguishes between internal dynamics, as stimulated by the presence of cities, and exogenous development pressures.

Nevertheless, the integration of rural areas into FUA is not necessarily an obvious development strategy. It can even intensify the difficulties that confront rural areas, as well as generating new problems. It also raises the spectre of a loss of distinguishing characteristics for rural areas. Hence, it is appropriate to ask if such integration should be a paradigm for rural development. The enthusiasm with which the ESDP stresses the advantages of rural-urban integration should not deflect attention from the question of what is the spatial field over which it is reasonable to expect integration processes to be effective. This is a key issue for this chapter, which first investigates whether quantitative measures that can be applied across Europe can be used to characterize the spatial extent of city-centred development processes. The chapter then explores the degree of socio-economic heterogeneity that exists amongst rural areas within city-regions. It concludes, using qualitative information gathered from case study investigations, with an examination of common issues that these rural areas face, which need to be taken into account in development policies for urban-centred regions.

**Recognizing Peri-Urbanization and Peri-Urban Areas**

The term 'peri-urbanization' does not appear in the text of the ESDP, which refers rather to 'urban surroundings' or the 'surrounding countryside', without providing a deeper definition. The word itself is not common in every language. In the German language, for example, the preferred term is 'sub-urbanization'. Yet the term is a regular visitor to the scientific literature, even if more in Latin than in Northern European countries, where it refers to rural areas that are subject to the influence of a nearby city or town (Dematteis, 1998). How far the peri-urban area extends is an issue of debate, with some interpreting it as little more than the urban fringe (e.g. Errington, 1994). In this chapter we take a broader view, as we focus on integration processes, which leads us to an understanding of the peri-urban zone that is closer to that of an urban commuter catchment. Thus, peri-urban areas are the hinterlands of cities. Hence, while the term evokes images of the physical expansion of built-up areas through the absorption of additional land on the urban periphery (Evert, 2000), the interpretation used here extends to less geographically immediate urban impacts. Peri-urban zones are indeed affected by urban sprawl, so issues of land consumption confront them, as do attempts to use spatial planning to control the spread of built-up areas, but zones impacted on by urban spread effects do not encompass the totality of areas that are deeply integrated into city-centred regions.

Peri-urban areas are zones for which the ESDP underlines a necessity to pay particular attention, in order to ensure the sustainability of city-region development. Yet much of these zones have a rural character, so they are distinguished from core cities as well as from suburbs. Thus, peri-urbanization is not incompatible with continued agricultural activity, but it does involve rural economic and social change. This is recognized in the ESDP, for even when this document is not referring explicitly to peri-urbanization, it deals with relations between cities and rural hinterlands within urban-centred regions.

In this context, although peri-urbanization attenuates traditional oppositions between cities and rural areas, the dual nature of peri-urban areas makes it difficult to distinguish them as a separate category of space, in the sense of being different from urban and rural areas. It follows perhaps that peri-urban areas are barely mentioned in legislative texts and, even when they are, tend to be seen simply as intermediate zones, without legal status.[1] Generally speaking, in research or in drafting territorial-planning policies, peri-urban areas (or their equivalent) may occasionally be identified in territorial typologies or other scientific documents, but their definition results more from the extension of urban categories than from true awareness of the dual nature of these areas.

One pending difficulty in dealing with these issues is that peri-urban areas are still an emerging category, whose importance is difficult to assess. A broader recognition of this category needs methods that seek first to locate the spatial extent of peri-urban processes, and then show possible disparities between peri-urban and other areas, in order that development policies can be adapted to different contexts.

---

1   For example, the term peri-urban appears for the first time in French legislation in the 'Loi d'orientation pour l'aménagement et le développement durable du territoire', which translates as 'Outline law on planning and sustainable development of the territory' (LOADDT, 1999, pp.9515-9527), whose Article 6 specifies that '... the regional territorial-planning and development plan defines notably the main goals concerning the choice of installation sites for major projects, infrastructure and general public services that should contribute to ... the harmonious development of urban, peri-urban and rural territories'.

## The Importance of Peri-Urban Areas in City-Regions

As peri-urban areas are considered to be predominantly rural areas located in city- or urban-centred regions, methods to show the spatial extension of peri-urban processes need to identify, as a first step, cities that act as the foci of urban regions (e.g. Coombes, *et al.*, 1982). It is then necessary to delimit areas under the influence of these centres; in other words to demarcate zones that comprise a functional urban region. Finally, it is necessary to distinguish, within these regions, peri-urban from urban areas, with this distinction not necessarily reducible to central cities against the rest.

### Core Cities as Centres of City-Regions

According commonly used approaches toward to urban region delimitation (e.g. Cheshire and Hay, 1986; Brunet, 1989; Cattan *et al.*, 1994; Gödecke-Stellmann, 1995; ESPON, 2004) core cities are the focal point of functional urban regions, which exert most influence over areas that comprise the FUA. In general terms, cities are characterized by a concentration of people, activities, capital and buildings. Their functions are characterized by flows of people, goods, energy, information and money. Their development is manifest through urban expansion beyond their limits, with the development of inter-relationships beyond their built-up area resulting in complex systems, which see the spheres of influence of different cities overlap.

One difficulty in using official data sources to apply this approach in Europe is the fact that there is wide variety in the definition and understanding of what constitutes a city (Cattan *et al.*, 1992). Even if there was uniformity, such cities have multi-faceted impacts, so their spheres of influence are difficult to delimit precisely or to rank in a hierarchy (Coombes, 2000). In some countries, cities are defined by administrative standards that confer a specific legal status, as with the German Städte. In Germany, some municipalities are part of upper level administrative units, known as *Kreise* (districts), but those municipalities that have been granted city status, due to their urban character, their number of inhabitants, their settlement pattern and their cultural and economic importance, constitute a district in their own right. In other countries, cities are defined mainly or even exclusively in statistical terms. For example, in Italy and in Spain, with the exception of Madrid and Barcelona, which are legally specified entities, cities are defined by a population threshold (more than 10 000 inhabitants). Offering another approach, some national statistical offices provide the definition for cities, for which they commonly draw on population size and settlement structure. This approach is used in France, where continuously built-up areas and the attributes of the population in these areas holds the key. In this scheme, several municipalities are commonly grouped together into so-called 'urban units', with the underlying philosophy for this approach resting on the United Nations urban area concept, which is based on distance between buildings, once allowance is given for areas of 'urban land-use' (e.g. sport arenas, parks, etc.) (Study Programme on European Spatial Planning, 2000).

A common characteristic of these definitions is that they do not take into account the potential influence, or capacity to stimulate development in nearby areas, of urban cores. Yet most typologies aim to rank European cities by their economic strength. They differentiate types of cities that act as development centres because their economies are well integrated into global networks and have developed high-technology activities (Study Programme on European Spatial Planning, 2000), yet without stressing influences on proximate zones. This is despite the fact that the most important cities should be centres of broad networks of inter-dependency, where centrifugal flows from these centres result in successive internal rings being integrated with the central city, thereby forming a metropolitan area (Precedo, 1988). This is not to say that no schemes for city delineation have recognized these 'spill-over' effects. In Germany, for example, the first attempts of the 1950s to define cities at the core of urban regions stressed not only their the physical extension beyond administrative borders but also their attraction as expressed through commuting (Duss, 1997), while some territorial typologies do take into account reciprocal relations between cities and rural communes (Perlik, 1999).

In fact, the importance of commuting, in a direction that reveals how cities act as 'centres', is a relevant criteria for evaluating city influence on nearby areas, since commuter flows take into account both spatial dimensions (flows usually decline with distance) and functional dimensions (integration into urban labour markets). From this perspective, city-regions identify areas from which commuter flows converge on particular cities (Coombes *et al.*, 1982). Using this approach, analysts can identify the relative importance of cities, by graphing a hierarchy of inter-connections between places, based on attractiveness of one city to another. Higher ranked cities in a nested hierarchy form the core of the largest city-regions.

*Locating City-Regions in ESPON*

This approach to locating the core and extent of city-regions may result in different rankings, in terms of city importance, than those obtained using the population size of continuous build-up areas. For instance, in France, according to INSEE (Le Jeannic, 1997; Bessy-Petry *et al.*, 2000), the Strasbourg urban area (a city with an extensive built periphery) was eleventh in the urban hierarchy in 1999, but ninth in the city-region hierarchy, due mainly to the importance of its 'periphery' (i.e. nearby areas influenced by the central city). Figure 2.1 shows the location of 'urban regions' in France, and the spatial extent of their respective cores.

The ESPON Project, in attempting to provide operational content for guidelines presented in the ESDP, located city-regions in Europe in order to study their spatial development and, in particular, the degree to which they reveal polycentric or monocentric settlement structures. In the ESPON study on polycentrism (ESPON, 2004), city-regions are defined as Functional Urban Areas (FUA), using the criteria that there are 'core municipalities', with more than 10 000 inhabitants, with 'fringe municipalities', so the total FUA population exceeds 20 000, with at least 50 per cent of journey-to-work trips taking place within the FUA.

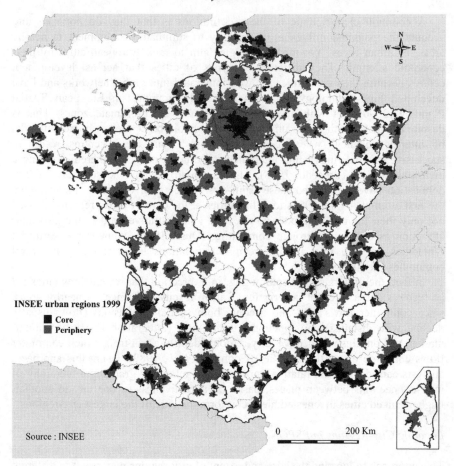

**Figure 2.1 INSEE 'Urban Regions' in France, 1999**

The ESPON analysis was based on existing population and commuter flow data for European countries or drew on national statistical office efforts to specify commuter catchment areas. To deal with differences in city definitions, and to make possible cross-national analysis using comparable data, NUTS5 territorial units were used.[2] Thus, core municipalities were NUTS5 units (or groupings of such units) with more than 10 000 inhabitants; irrespective of whether these are 'cities', according to national definitions. The main difficulty ESPON faced was the availability and comparability of commuting data, along with the different size

---

2  NUTS5 is the most detailed level in the nomenclature of statistical territorial units (NUTS) set up by Eurostat. The NUTS5 level corresponds to basic administrative units (equivalent to communes or municipalities) in most countries where this level has been established. In the UK, NUTS5 corresponds to electoral wards rather than administrative units (Eurostat, 2003).

of NUTS5 units in each country, as these can create bias in cross-national analysis. Commuting data were available at the NUTS5 level only for Austria, Belgium, Denmark, Finland, France, Germany, Luxembourg, Sweden, and Norway. National counterparts to the FUA definition exist in these countries, and also in the Czech Republic, Greece, Hungary, Italy, the Netherlands, the Slovak Republic, Slovenia, the United Kingdom and Switzerland. There was less cross-national comparability in other European countries, including the Iberia Peninsula and the Baltic Sea states. Despite these limitations, the ESPON exercise identified some important features. As shown in Figure 2.2 and Table 2.1, city-regions with a population of more than 100 000 inhabitants account for more than 50 per cent of Europe's population, with the highest population shares in some northern countries (Denmark and Sweden), as well as in Belgium and some Mediterranean countries (Italy and Spain). The lowest percentages are in alpine countries (Austria and Switzerland). The rather low percentage in the UK is due to the small size of travel-to-work areas that were used as FUA in order to provide valid counterparts to other national settings.

Across nations, core cities share distinctive urban features, such as high population densities, a concentration of workplaces and activities, and a predominance of built-up land-uses. Other municipalities share these features, whether they qualify as cities or not. By contrast, peri-urban areas exhibit less marked urban features, with rural characteristics retained over much of their territory. Given their deep penetration by urban-centred networks and flows, in order to identify the importance of peri-urban zones within European city-regions we need to identify those attributes that distinguish peri-urban from urban areas.

Scientists have long discussed what is a rural area, with scientific debate likely to continue indefinitely. There is nevertheless rather general agreement that these areas possess different land-uses, settlement structures and living conditions than cities, even if there is debate over how far these are specifically 'rural' characteristics (e.g. Bock, 2004). Attempts to capture rural populations and spaces while maintaining a material understanding of 'the rural' bring out diversity within the rural realm, the key dimensions of which can be based on intuition, theory or empirical regularities, which lead to the identification of explicit geographical categories (ESPON, 2004a). Linked to these ideas, spatial delineations of rural area-types assist in the geographical targeting of resources to areas with particular needs (e.g. Hannan and Commins, 1993; Malinen, 1995).

*Urban and Peri-Urban Areas Within City-Regions*

These distinctions indicate that there is diversity amongst rural areas, but this diversity is not a great help in defining what is a rural area as opposed to an urban area. In practice, identifying rural areas is often based on degrees of absence of 'urban features'. By contrast, the OECD has classified rural areas by their degree of access to major urban centres, thereby acknowledging that some areas are economically integrated into major urban centres while remaining rural in population density terms (OECD, 1993). According to this concept, the urban and the rural maintain close relationships within city-regions, with the ties that bind

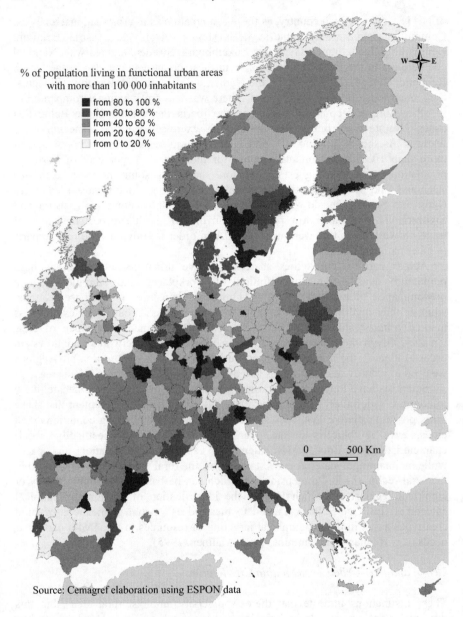

**Figure 2.2 Percentage of NUTS2 Populations in Functional Urban Areas of 100 000 or More Inhabitants**

**Table 2.1  The Demographic Importance of ESPON's Main Functional Urban Areas in European Countries**

| Country | FUA of more than 100 000 habitants | | |
|---|---|---|---|
| | Number | Population in 1999 | Share of total population |
| Austria | 5 | 2 215 925 | 27.4 |
| Belgium | 20 | 7 084 316 | 69.4 |
| Cyprus | 2 | 411 366 | 54.0 |
| Czech Republic | 21 | 5 476 652 | 53.3 |
| Denmark | 12 | 4 065 271 | 76.4 |
| Estonia | 2 | 635 300 | 44.1 |
| Finland | 9 | 2 781 790 | 53.9 |
| France | 78 | 35 305 341 | 60.2 |
| Germany | 90 | 42 658 368 | 52.4 |
| Greece | 5 | 5 298 175 | 50.3 |
| Hungary | 19 | 4 734 830 | 47.0 |
| Ireland | 2 | 1 200 600 | 32.1 |
| Italy | 119 | 36 797 371 | 61.9 |
| Latvia | 3 | 1 444 725 | 59.4 |
| Lithuania | 5 | 1 375 300 | 37.2 |
| Luxembourg | 2 | 259 901 | 60.2 |
| Malta | 1 | 380 000 | 100.0 |
| Netherlands | 24 | 8 992 904 | 56.9 |
| Norway | 9 | 2 393 161 | 53.9 |
| Poland | 46 | 19 378 376 | 50.1 |
| Portugal | 6 | 4 355 114 | 45.1 |
| Slovak Republic | 13 | 2 798 438 | 51.9 |
| Slovenia | 3 | 910 216 | 45.9 |
| Spain | 62 | 26 026 109 | 67.1 |
| Sweden | 24 | 6 726 491 | 76.0 |
| Switzerland | 9 | 2 929 253 | 41.0 |
| United Kingdom | 61 | 25 376 285 | 42.7 |
| Total | 658 | 250 612 558 | 54.5 |

Source:   Cemagref elaboration using ESPON data.

them definable through structural associations and functional relations. By structural associations the OECD draws attention to relatively stable relationships, as with location, land-use, settlement structure, population distribution or administrative arrangements. Functional relations, by contrast, refer to production, consumption and communication flows that express the way the physical environment is utilized in activities that reinforce links within city-regions; so they are more likely to change over time (ESPON, 2004a).

When seeking to distinguish rural from urban, a focus on structural or functional properties offers a different perspective from that derived if distinctions draw on 'rurality' related concepts. Yet there is no doubt that land-use, settlement patterns and population size, as well as commuter flows, do define different types of geographical area; as seen in the share of the land surface that is open space. But there is an essential difficulty in applying these criteria for the whole Europe, given the (non-) availability and comparability of data to represent these criteria.

ESPON delimitations of FUA were affected by differences in the availability and comparability of commuting data, and thus did not try to explore differences within city-regions or across nations. In order to explore these differences, work associated with the NEWRUR Programme examined the situation in France, Germany, Greece, Spain and the UK.[3] It soon became clear that the absence of comparable data across these nations limited attempts to provide a common approach to delimiting the scale of city-regions, as well as diversity within these regions. For analyses across nations, therefore, the investigation was forced to focus on structural properties that could be expressed with the help of population figures at the NUTS5 level, with comparable data consisting of little more than population censuses data. As with ESPON, in order to define city-regions, the analysis started by specifying what are core cities, with these units defined by amalgamating NUTS5 units using population thresholds for continuous built-up areas. These thresholds were fixed for each country, in line with current scientific thinking, so as to reflect the level of population 'required' to impact on nearby zones and thus lead to peri-urbanization. Other settlement categories were then defined by applying population density thresholds, which were fixed for each country separately, with this density measure combined with contiguity criteria (Table 2.2). The first areas to be so identified were 'suburban' units, which are contiguous to core units and can be regarded as more urban than rural, due to their high population density, so they are extensions of the central city built-up zone. Peri-urban rural units were then identified, by specifiying a lower population density, whiel insisting on contiguity with core cities or suburban areas.

In Spain and Greece, these measures of peri-urban areas were found to be broadly acceptable, as density differences appeared to be linked to distance from core cities, as well other structural properties, such as different administrative roles. But in France, Germany and the UK, population density differences appear

---

3 Funded by the European Commission's initiative on: 'Sustainable agriculture, fisheries and forestry and integrated development of rural areas including mountain areas', the Programme's title was 'Urban pressure on rural areas: mutations and dynamics of peri-urban rural processes'. Reports are found at http://newrur.grenoble.cemagref.fr/.

to be due to important factors in addition to distance from a core city. Thus, other measures are required to locate peri-urban areas. In seeking better integration measures attention was directed at functional relationships that could be described with available data, such as population flows or travel to work displacements. The availability of such data would have provided more accurate peri-urban areas in Spain or in Greece.

**Table 2.2 Indicators for Identifying Urban, Suburban and Peri-Urban Areas in Five European Countries**

|  | Type of NUTS5 area | Country | | | | |
|---|---|---|---|---|---|---|
|  |  | France | Germany | Greece | Spain | UK |
| Criteria | Core city | More than 20 000 inhabitants (1999) or more than 10 000 at a density above 300 inhab/km² | More than 80 000 inhabitants (1999) | More than 10 000 inhabitants (1991) | More than 30 000 inhabitants (1991) or more than 10,000 at a density above 150 inhab/km² | Density above 1 500 inhab/km² (1991), for built-up area of at least 50 000 inhabitants |
|  |  | Contiguous to a core city and | | | | |
|  | Suburban area | Density above 200/km² | Density above 500/km² | 2 000 to 10 000 inhabitants | Density above 150/km² | Density above 400/km² |
|  |  | Contiguous to a core city or its suburban units and | | | | |
|  | Peri-urban area | Density above 75/km² and falls within the employment catchment of the relevant core city | Saw an increase in population from 1996 to 1999 | Density above 50/km² | Density above 50/km² | Density above 150/km² and falls within the travel-to-work area of the relevant core city |
| Share in a nation's population | Core and suburban units | 55.4 * | 45.9 | 58.3 | 65.0 | 50.5 |
|  | Peri-urban units | 7.0 * | 13.2 | 27.2 | 9.6 | 4.7 |
|  | City-regions | 62.4 * | 59.1 | 75.5 | 74.6 | 51.2 |

Note: * For city-regions with city and suburban populations over 50 000.

In spite of this limitation, this method made it possible to specify peri-urban areas in each country and suggests their relative importance within city-regions, as shown in Table 2.2. The relative weight of peri-urban areas in NUTS2 regions in the five countries investigated varies from 0 per cent to 62 per cent. The lowest values are in 'capital city' regions, in regions dominated by a major city or in regions with a particularly low urban population (e.g. Basse-Normandie in France, Niederbayern an Oberfranken in Germany, Castilla-la Mancha in Spain, Cumbria and Lincolnshire in UK, etc.). The highest values are for places like Alsace in France, Leipzig in Germany, Kentriki Makedonia in Greece and Cantabria in Spain, where the population share in urban areas is rather high, although it is also substantial in regions where urban zones are less important demographically (e.g. Oberpfalz and Gießen in Germany, and Sterea Ellada and Peloponisos in Greece).

The systematic availability of data on population density makes it possible to plot percentage shares in different 'environments', while adjustment of thresholds can be used to make allowance for differences in the size of NUTS5 units. However, the method is very simplistic. Other methods of delineating peri-urban areas can be imagined. For example, the recent update of Corine Landcover data describes diverse types of land-use, including 'artificial surfaces', which makes it possible to locate potential rural areas (European Environment Agency, 2005).

**Diversity in Peri-Urban Areas**

The territorial analyses carried out in the ESPON Project did not go beyond delimiting FUA. The internal operations of FUA were not studied in detail. Yet, peri-urbanization has tangible effects, as seen in the concentration of population growth in peri-urban zones. This is fairly general across Europe, where many cities are stagnating or declining demographically, whereas surrounding areas receive new residents from nearby cities or even from farther away, which indicates the attractiveness of these areas within city-regions (e.g. Schmied, 2005).

Although demographic growth in city-regions is often focused on peri-urban zones, the demographic effects of urban expansion do not necessarily reveal the diversity of influences exerted by cities on nearby areas. This encourages more depth in the investigation of what plays a leading role in rural-urban integration, and how these affect peri-urban areas in different ways. However, focusing on integration should not lead to the neglect of factors that create differences between peri-urban areas. To formulate and implement development objectives more efficiently in city-regions, integration and other factors need to be taken together in order to understand development diversity within these areas.

*Differences in Rural-Urban Integration*

Integration approaches focus on linkages within a system that makes actions better conceptualized as joint activities, or where linkages between one and another is so strong that action by one has a major impact on the other. Integration approaches have been used to identify coherent geographical boundaries for human 'systems'

(Coombes, 2000). Commonly, such investigations are based on linkages that consolidate – or alleviate – cohesion within a system. Assuming city-regions are systems comprising urban as well as peri-urban units, integration approaches help identify factors playing a leading role in system cohesion.

Current statistical investigations of rural-urban integration in city-regions privilege linkages associated with travel-to-work displacements. This is a consequence of two main factors: the increasing disassociation of places of residence and work; and the concentration of jobs in development poles. There are differences in the intensity of these phenomena, which highlight diversity in rural-urban integration. This has been aknowledged in some official typologies, as with the Urban Regions model of the German Federal Office for Building and Regional Planning (BBR, 2000), which distinguishes core cities, inner commuter catchment areas, and finally outer commuter catchment areas, according to the share of the resident workforce that commutes to the core city. In France, the INSEE Urban Areas typology illustrates another example (Le Jeannic, 1997) with different types of peri-urban 'commune' recognized, depending upon whether workers are drawn principally to a single city (or to communes whose resident workforces are attracted to this city), which are referred to as monopolarized communes, or are drawn to several urban centres (so-called multipolarized communes).

In fact, rural-urban integration can be identified through various functional criteria, which express tendencies toward the growth or decline of city influence on rural areas. In the Study Programme on European Spatial Planning, for example, research focused on conceptualizations of spatial differentiation that favour or hinder regional integration, in the sense of drawing closer or moving away from the over-arching goal of a balanced, polycentric European territory. These criteria are linked to economic development, social integration and cultural criteria (Study Programme on European Spatial Planning, 2000).

In terms of integration differences, the manner this has been approached in our work is by way of spatial and function perspectives. Spatially, it is recognized that integration is made easier by the proximity of rural areas to cities. Functionally, integration is seen to emerge from the development of complementarities between rural zones and cities, whether this be in job distributions, household services, land-use allocation or in other domains. Yet local resistance can mitigate rural-urban integration, with a willingness to cooperate by no means assured in rural communities (e.g. Woods, 2005).

According to these principles, NEWRUR research on rural-urban integration moved beyond population density measures to explore a wide range of interaction indicators at the city-region level. The intention was to assess the intensity of the urban influence on rural areas, using quantitative indices of spatial, functional or even institutional links. Besides spatial links, which can be illustrated by distance to core city indices, the measures that were used were drawn from different spheres, in order to capture the most prominent links with cities. They also had to differ from country to country, owing to the uneven availability of data across, as well as due to differences in what was considered worthy of deeper investigation in each country. For example, in France, attention was paid to institutional relations (inter-municipal co-operation) or service supply links that can lead to associations within peri-urban areas and in some cases lead to conflict with central cities (see Chapter Four). In Spain, the capacity to answer the social expectations of residents is important, which is illustrated by concerns over distances to hospitals and the availability of residences for old people. In the UK, indices focused on journey to work flows with main or secondary employment centres. The synthesis of different measures into a general quantitative measure, led to an index that expressed spatial, functional and/or institutional integration, which was used to identify different of types of rural area, according to the intensity of their links with cities (on a similar idea, using different methods, see Blunden *et al.*, 1998).

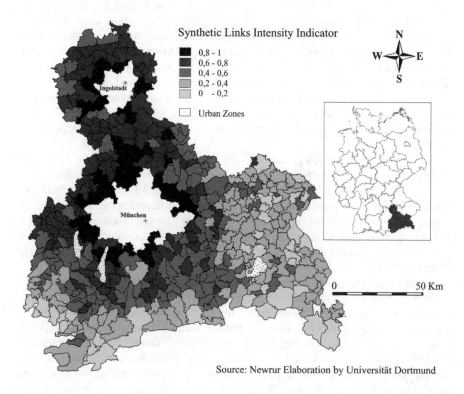

Source: Newrur Elaboration by Universität Dortmund

**Figure 2.3 Peri-Urbanization Gradients in Upper Bavaria**

In applying these ideas, a set of NUTS2 regions were selected that reveal differences in peri-urbanization, as expressed by the share of peri-urban areas in regional populations.[4] For these regions, the mapping of index scores revealed first and foremost a clear peri-urbanization gradient, in that the intensity of linkages

---

4   These were: Rhône-Alpes and Provence-Alpes-Côte d'Azur in France, Oberbayern in Germany, Andalusia and Murcia in Spain, and Bedfordshire-Hertfordshire and East Anglia in the UK.

generally decreases with distance from cities, although this does not mean that integration processes are not underway in rural areas that appear isolated from cities. Even so, the index measures distinguished between 'strictly peri-urban areas' and other peri-urban areas, according to the strength of linkages with cities. Figure 2.3 shows how rural-urban interaction measures decrease with distance from Munich, while Figure 2.4 identifies similar relationships for Rhône-Alpes and Provence-Alpes-Côte d'Azur.

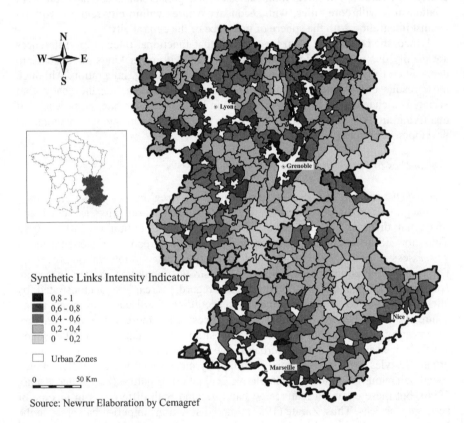

Source: Newrur Elaboration by Cemagref

**Figure 2.4 Peri-Urbanization Gradients in Rhône-Alpes and Provence-Alpes-Côte d'Azur**

Besides distance to cities, economic factors prevailed in most regions in distinguishing between different types of peri-urban zone. The clearest opposition arose between 'dormitory' peri-urban communities and 'secondary employment poles', where links with core cities are partly mitigated by the fact they play a role as local employment centres. Other kinds of linkage did not appear to play the same role as economic linkages. For example, in France, service supply links introduce differences amongst rural zones that are not similar to those as a result of economic relations with city centres, due to the fact that these relationships primarily exist between nearby communities, not core cities. In the same way, the development of inter-municipal cooperation structures, which group rural communities to organize current services and so regulating urban development effects, can be interpreted as an attempt to mitigate dependency towards core cities. In contrast with France, in the Spanish regions, agricultural areas have traditionally important service supply links with key urban centres. Then again, in Greece, analysis of rural-urban integration indicates that peri-urban zones have different relationships with core cities, with secondary centres within city-regions working against integration with the sphere of influence of the central city.

Hence, the conclusion from this work is that functional integration differences are mainly due to economic relations with cities, with other kinds of relationship deepening or diversifying, or even in some cases mitigating, integration. Although these results only relate to a particular set of European regions, they show that variety is a characteristic of peri-urbanization, and so demonstrate the necessity of qualifying any vision of uniformity over peri-urban zones in regional planning or development policies.

*Approaches to Peri-Urban Diversity*

The ESDP recognizes heterogeneity within city-regions by specifying that: '... integrated urban development strategies should be sensitive to social and functional diversity' (Commission of the European Communities, 1999, §88). Thus, the ESDP encourages deeper analysis of peri-urban heterogeneity, along with assessments of the degree to which diversity is linked to integration processes. For example, although peri-urban areas see strong population growth in certain countries, they are quite different in this regard, as can be demonstrated by allocating expected population increases according to distance from an urban centre, household income or real estate prices; with these differences in turn accentuating imbalance in the social composition of populations (Gofette-Nagot, 1995). In the same way, the peri-urban fringe may be characterized by increasingly urban lifestyles, by stronger social mobility, by more social variety, differentiated social behaviour or a more contemplative evaluation of nature (Zárate and Rubio, 1990), but these common features go hand in hand with differences in income or access to services. Thus, Zárate (1984) found distinctions in peri-urban areas in the kind of connections residents had with a central city, with clear differences between those on main access routes and those in scattered settlement systems.

From their part, European or national typologies of rural space tend to emphasize general socio-economic characteristics rather than relations with cities. With regard to development perspectives, the working document of the European Commission's *Europe 2000* distinguished several types of rural spaces, according to their main economic activities ortheir remoteness, with rural spaces next to large cities distinguished from others (Commission of the European Communities, 1994). In the UK, some classifications are based on rural linkages to cities (Coombes *et al.*, 1982), but other typologies, as for example the Office of National Statistics's classification of local government districts, do not take this into account explicitly (Wallace and Denham, 1996; Bailey *et al.*, 1999). Hence, given that differences between peri-urban areas can be smoothed out in general rural typologies, there is no immediate answer to the question, is heterogeneity in peri-urban zones explained more by uneven integration with cities or by factors that concern both peri-urban and other rural areas? This is why, in the research on which this chapter is based, this question was explored for a set of NUTS2 regions using a case study approach.

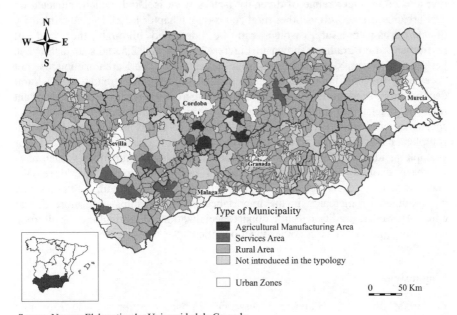

Source: Newrur Elaboration by Universidad de Granada

**Figure 2.5 Peri-Urban Diversity in Andalusia and Murcia**

In Rhône-Alpes and Provence-Alpes-Côte d'Azur, a cluster analysis of socio-economic variables revealed that tourism, in-migration, type of economic activity, rural-urban job balance and access to everyday services, play a leading role in rural heterogeneity. Yet, for peri-urban areas, economic differences are less marked than in other rural areas, so they are less important in distinguishing between peri-urban places. In Upper Bavaria, the main distinction within peri-urban areas is between 'growing municipalities' and peri-urban zones that do not enjoy the same opportunities. In Andalucia and Murcia, economic differences are notable, even for places with similar linkages with cities, although most peri-urban municipalities fall into a 'rural areas' cluster, as shown in Figure 2.5, where there is little economic diversity. As for the East of England, what emerged here is a pattern of broad consistency for economic conditions in rural areas, save for the punctuation of a few places that are most distant from cities or are most proximate to London (e.g. Henderson and Hoggart, 2003). Indeed, in a cluster analysis of socio-economic conditions in this region, the main diversity factors across NUTS5 zones related to income/occupation, with more than half of peri-urban wards falling into the cluster defined as 'higher income zone'.

These exercises show that in the French and German regions diversity in peri-urban as well in other rural areas is smoothed by the strength of linkages with urban centres. Peri-urban diversity is revealed first and foremost by a peri-urbanization gradient, then by secondary factors. In the other nations investigated, a wider range of diversity factors were isolated, which impacted on peri-urban zones as well as other rural zones (e.g. Chapter Five). Here the intensity of peri-urban pressure, which can be identified through the kind of peri-urbanization gradient shown in Figures 2.3, 2.4 and 2.5, fails to distinguish between different types of peri-urban area. Thus, peri-urbanization is not necessarily the most important diversity factor, as more global trends, such as rural depopulation, tourism demand or agricultural restructuring, play a more important role. In fact, peri-urban diversity may reveal differences in functional specialization within peri-urban areas. Some specializations express functional complementarities with core cities, that peri-urban processes may reinforce, as for example residential functions, or even recreation or leisure functions, which have not been evoked yet. As many residential peri-urban zones have developed manufacturing or even high-tech activities, complementarities with core cities may be smoothed by increases in the importance of relationships with more distant cities. However, traditional functional specialization in agriculture or in food processing may coexist with other functionalities.

## Conclusion

Peri-urban diversity, whatever its essence, embodies differences in economic context, which leads to uneven economic growth and the unequal influence of forces other than those which strenghten ties within city-regions. For these zones, according to the ESDP, besides controlling a city's physical expansion, the main issues are ensuring social and functional heterogeneity, managing the ecosystem effectively and efficiently, improving transport accessibility using environmentally friendly options and ensuring natural and cultural heritage conservation and development. Due to their diversity, peri-urban communities are not equally placed to face these issues, nor do they have the same opportunities to deal with them. Thus, tensions which ensue can be contained and reduced more or less easily from place to place.

In terms of these tensions, for the regions investigated for this research, there is a common tendency for jobs to be increasingly separated from residences, which is increases transport problems and poses questions for longer term environmental sustainability. Yet different tensions exist over securing non-farm jobs, except in service industries catering for local housing growth, for while there are new employment centres on the periphery of some city-regions, as well as near major transport inter-connections, such employment nodes are localized, not dispersed. In four of the nations investigated (France, Germany, Greece and Spain) municipal competition for new jobs is leading to the indiscriminate construction of industrial parks within the peri-urban zone, many of which have a multitude of 'unfilled' places.

Within peri-urban areas there is also a common tendency to attract socially selective in-migrant groups, namely younger households and those with families, but pressure for housing is forcing up prices, such that candidates for residence are now finding it difficult to afford homes. The tendency toward social exclusivity is then intensified by a shortage of social housing. Adding to shortfalls in low-cost housing throughout city-regions, this tendency is reinforced by high demand for individual housing units (rather than apartments), with this housing type most commonly associated with peri-urban and rural places of residence. Added to which, while fiscal deficiencies encourage small municipalities to favour housing growth in order to enhance their tax revenues (e.g. Chapter Three), with recently arrived residents commonly opposing new housing developments, especially if these are apartments.

In terms of infrastructure provision, population expansion in peri-urban areas helps maintain or even expand service provision, although where the source of development pressure comes not from the immediate city but from more 'global' influences, such as tourism or producing for export, enhancements in service levels can be directed more toward temporary visitors rather than the permanent population (e.g. Chapter Five). Then again, even where the source of dynamism for demographic, economic and housing change in peri-urban zones originate in the core city of a city-region, there is often imbalance in service provision as population needs outstrip available infrastructure and service capacities. Thus, there is a growing tendency for the core city to be relied on for services (e.g. Chapter Six), with some mitigation where a city-region includes subsidiary centres. All to commonly, this central city reliance comes with a heavy dependence on private transport options. Scattered and small populations work against effective public transport provision.

These tensions raise questions about the prospect of local government mitigating either the outcomes or the processes that increase imbalances within peri-urban zones. The actions of local or other governmental agencies are restricted by a lack of effective coordinating governance arrangements across institutions in a city-region, as well by small municipalities being associated with poor fiscal resources, given the demands for services that are increasingly being placed on them. Well coordinated development strategies that extend beyond municipal or other administrative boundaries would be one means of dealing more efficiently and effectively with uneven tensions in city-regions. In this context, peri-urban diversity is not simply a matter of uneven integration with urban centres, but also arises from issues that link rural to rural and rural to urban. This poses the question, would it be possible to develop analytical methods to assess peri-urban diversity that go beyond descriptive indicators to encompass issue-oriented processes?

## References

Antikainen, G. and Vartiainen, P. (2002) Finnish districts and regional differentiation, *Fennia*, 180, 1-2
Bailey, S., Charlton, J., Dollamore, G., Fitzpatrick, J. (1999) *The ONS Classification of Local and Health Authorities in Great-Britain: Revised for Authorities in 1999*, The Stationery Office, London
BBR (2000) *Urban Development and Urban Policy in Germany: An Overview*, Bundesamt für Bauwesen und Raumordnung, Berichete Band 6, Bonn
Berger, M., Fruit, J.P., Plet, M.C. and Robic, M.C. (1980) Rurbanisation et analyse des espaces ruraux périurbains, *Espace Géographique*, 9, 303-313
Bessy-Petry, P., Julien, P. and Royer, J.F. (2000) De nouveaux périmètres urbains pour la France de l'an 2000, *Données Urbaines*, 1, 173-185
Blunden, J.R., Pryce, W.T.R. and Dreyer, P. (1998) The classification of rural areas in the European context: an exploration of a typology using neural network applications, *Regional Studies*, 32, 149-160
Bock, B.B. (2004) It still matters where you live: rural women's employment throughout Europe, in H. Buller and K. Hoggart (eds.) *Women in the European Countryside*, Ashgate, Aldershot, 14-41
Brunet, R. (1989) *Les villes européennes*, DATAR, Paris
Cattan, N., Pumain, D., Rozenblat, C. and Saint-Julien, T. (1992) *Le concept statistique de la ville en Europe*, Eurostat, Luxembourg
Cattan N., Pumain, D., Saint-Julien, T. (1994) *Le système des villes européennes*, Anthropos, Paris
Cheshire, P.C. and Hay, D. (1986) The development of the European urban system 1971-1981, in H.J. Ewers, J.B. Goddard and H. Matzerath (eds.) *The Future of the Metropolis: Berlin, London, Paris, New York - Economic Aspects*, de Gruyter, Berlin, 149-169
Coll, J.L., Grasland, Cl., Pumain, D. and Saint-Julien, T. (2000) *Préfiguration de l'observatoire en réseau de laménagement du territoire européen - rapport remis à la Datar par le Point Focal Français*, at www.cybergeo.presse.fr
Commission of the European Communities (1994) *Europe 2000: Co-operation for European Territorial Development*, Office for Official Publications of the European Communities, Luxembourg
Commission of the European Communities (1999) *ESDP – European Spatial Development Perspective: Towards Balanced and Sustainable Development of the Territory of the European Union*, Office for Official Publications of the European Communities, Luxembourg
Coombes, M.G. (2000) Defining locality boundaries with synthetic data, *Environment and Planning*, A32, 1499-1518
Coombes, M.G., Dixon, J.S., Goddard, J.B., Openshaw, S. and Taylor, P.J. (1982) Functional regions for the population census of Great Britain, in D.T. Herbert and R.J. Johnston (eds.) *Geography and the Urban Environment: Volume Five*, Wiley, Chichester, 63-112
Dematteis, G. (1998) Suburbanización y periurbanización – ciudades anglosajonas y ciudades lartinas, in *La ciudad dispersa, suburbanización y nuevas periferas*, Monclús, Barcelona
Duss, R. (1997) Die stadtregion - ihre bedeutung für die regionale zusammenarbeit in den städtische regione, *Stadtforschung und Statistik*, 1/97, 19-31
Errington, A.J. (1994) The peri-urban fringe: Europe's forgotten rural areas, *Journal of Rural Studies*, 10, 367-375

ESPON (2004) *Potentials for Polycentric Development in Europe*, ESPON Project 1.1.1 Final Report, at www.espon.lu
ESPON (2004a) *Urban-Rural Relations in Europe*, Espon Project 1.1.2 Final Report, at www.espon.lu
European Environment Agency (2005) *Corine Land Cover 2000*, at http://www.eea.eu.int/
EUROSTAT (2003) *The Eurostat Nomenclature of Territorial Units for Statistics*, at http://europa.eu.int/comm/eurostat
Evert, K.-J. (2000) *Lexikon Landschafts- und Stadtplanung*, Springer-Verlag, Heidelberg
Faludi, A. and Waterhout, B. (2002) *The Making of the European Spatial Development Perspective: No Masterplan*, Routledge, London
Gödecke-Stellmann, J. (1995) Auf dem weg zu einer neuabgrenzung der stadtregionen, *Stadtforschung und Statistik*, 1/95, 64-71
Goffette-Nagot F. (1995) *Périurbanisation et composition socio-démographique des communes rurales*, in Colloque: Territoires ruraux et formations, Etablissement d'Enseignement Supérieur Agronomique de Dijon, Département Sciences de la Formation et de la Communication, Dijon
Hannan, D.F. and Commins, P. (1993) *Factors Affecting Land Availability for Forestry*, Economic and Social Research Insitute Monograph 76, Dublin
Henderson, S.R. and Hoggart, K. (2003) Ruralities and gender divisions of labour in Eastern England, *Sociologia Ruralis*, 43, 349-378
Hilal, M. and Schmitt, B. (1997) Les espaces ruraux : une nouvelle définition daprès les relations villes-campagnes, *INRA - Sciences Sociales*, 5, 1-4
Jaillet, M.C. and Jalabert, G. (1982) La production de l'espace urbain périphérique, *Revue Géographique des Pyrénées et du Sud-Ouest*, 53(1), 7-26
Le Jeannic, T. (1997) Trente ans de périurbanisation: extension et dilution des villes, *Economie et Statistique*, 307, Septembre, 21-41
LOADDT (1999) Loi d'Orientation pour l'aménagement et le développement durable du territoire du 25 juin 1999, *Journal Officiel de la République Française*, 29 Juin, 9515 sq
Malinen, P. (1995) *Rural Area Typologies in Finland*, paper presented at the LEADER Workshop 'Typology of European Rural Areas', 2-5 November, Luxembourg
OECD (1993) *What Future for Our Countryside? A Rural Development Policy*, Organization for Economic Co-operation and Development, Paris
OECD (2002) *Redefining Territories*, Organization for Economic Co-operation and Development, Paris
Perlik, M. (1999) Processus de périurbanisation dans les villes des Alpes, *Revue de Géographie Alpine*, 87, 143-151
Precedo, L. (1988) *La red urbana*, Sintesis, Madrid
Pumain D. and Saint-Julien T. (1996, coord.) *European Urban Network*, John Libbey Eurotext Limited, Montrouge
Prost, B. (1993) Aux marges du système urbain: les espaces flous et leur évolution, *Méditerranées*, 1/2, 37-40
Schmied, D. (2005) Incomers and locals in the European countryside, in D. Schmied (ed.) *Winning and Losing: The Changing Geography of Europe's Rural Areas*, Ashgate, Aldershot, 141-166
Steinberg, J. (1991) Les formes de la périurbanisation et leur dynamique, in *La périurbanisation en France*, SEDES , Paris, 59-85
Study Programme on European Spatial Planning (2000) *Final Report*, in www.mcrit.com/SPESP
Wallace M. and Denham C. (1996) *The ONS Classification of Local and Health Authorities in Great-Britain*, Her Majesty's Stationery Office, London

Woods, M. (2005) *Contesting Rurality: Politics in the British Countryside*, Ashgate, Aldershot

Zárate, A. (1984) *El mosaico urbano: organización interna y vida en las ciudades*, Editorial Cincel, Madrid

Zárate, A. and Rubio, M.T. (1990) *Análisis de la ciudad: Espacio heredado, objetivo y percibido*, Guía didáctica, Universidad Nacional de Educación a Distancia, Madrid

Chapter 3

# Commuter Belt Turbulence in a Dynamic Region: The Case of the Munich City-Region

Claudia Kraemer[1]

## Introduction

A large German newspaper describes the 'success story of the new village' as '... the change from a purely agricultural village without a future into a rural community where something new has evolved: a mixed society, where a coexistence of different professions, generations and lifestyles has become understood' (*Frankfurter Allgemeine Sonntagszeitung*, 2003). According to this article, the village is 'panic-proof' and far from dying; it is also said to be proof that Germany has finally been urbanized completely. There is hardly a place where this becomes more evident than in large city hinterlands; in their peri-urban areas. Here, urban pressure and dynamics accelerate change and challenge inhabitants oft quoted self-conceptions about a liveable rural area that is striving to preserve a long-faded identity. Whereas cities often remain the 'symbolic centre' of a region, the new hinterland is a young and sometimes complex area with self-confident political actors (Aring, 2004).

The European Spatial Development Perspective (ESDP) seeks to overcome outdated dualisms between city and countryside as a spatial development guideline in order to promote an integrated perception of city-regions as 'functional entities' in spatial planning (Commission of the European Communities, 1999). This notion is based on the assumption that cities are not only continuing to grow physically beyond their former limits – which authorities should be containing by spatial planning interventions – but also that functional linkages between cities and their hinterlands are intensifying – which should to be built on so as to balance and share costs and burdens within city-regions (Commission of the European Communities, 1999, p.19).

German agglomerations have gone through decades of more or less intensive and varying types of 'suburbanization', thus integrating ever more distant 'rural'

---

[1] This chapter is based on the University of Dortmund's contribution to the NEWRUR project on 'Urban pressure on rural areas', which was led by Professor Volker Kreibich.

areas into their spheres of influence. This chapter focuses on those areas that are generally defined as rural but are close to urban centres. It examines how far the integration of city and countryside meets the objectives that are specified in the ESDP. In order to do this, the chapter focuses on the peri-urban belt of Munich in the south of Germany. In this peri-urban belt the dualism between rural and urban seems to be vanishing in favour of an intermingling of the two. Internal and external 'pressures' on these intermediate areas will be identified and their consequences highlighted, in order to show how these are working toward or against the hinterland's integration into the city-region. As one conclusion, the chapter will explore whether the trends identified can justify a conservationist approach, as pursued by those who seek to protect existing rural character against urban pressures, or whether it is valid to argue for a stronger and more positively promoted integration of peri-urban areas into city-regions.

## Perceptions of Urban and Rural as Spatial Categories

City administrative boundaries have long ceased to reflect the extent of urban agglomerations and their spheres of influence. Perception of the intensity and nature of urban influence can be interpreted from the statistical and analytical models that have been used to delimit city-regions and urban agglomerations, with three basic approaches used in the spatial analysis of urban spread processes in Germany. All three have share a 'negative' definition of rural areas, for they see them as residual categories, which reflects practical difficulties in grasping the complex concepts of 'urban' and 'rural' with a few variables. The Federal Statistical Office's (BBR) model of *'Siedlungsstrukturelle Gebietstypen'* ('structural area types') is based on the simple criteria of the size and density of a municipality. It depicts core cities, highly 'densified', 'densified' and rural counties within agglomerated areas. 'Urban regions' (*Stadtregionen*), which are based on the 1950s model of Olaf Boustedt (e.g. 1970), are delimited according to the additional functional criterion of the intensity of commuter flows (Göddecke-Stellmann and Kuhlmann, 2000). German 'conurbations' (*Verdichtungsräume*) in turn are units that were originally identified to assess problems of 'over-densification' in urban areas. The relevant criteria are an above-average population density (inhabitants per km$^2$ in residential areas), an above-average share of residential and transportation areas, and a minimum of 150 000 inhabitants within a contiguous territory. These conurbations are the basis for a more accurate delimitation of planning categories in state and regional development plans and programmes.[2] This scheme applies to Bavaria, where conurbations – with inner and outer parts recognized – are separated from rural areas of a number of different kinds, with the category 'rural areas around large conurbations' (*Ländliche Teilräume im Umfeld großer Verdichtungsräume*), or peri-urban areas, constituting one rural category.

---

2   These planning categories are the basis for the formulation of specific planning and development objectives and guidelines.

According to the Bavaria Regional Development Programme (LEP) these areas:

- are exposed to the strong influence of an urban agglomeration
- have had significant increases in the number of inhabitants and jobs
- are exposed to land-use pressures
- are beginning to develop higher densities because of disproportionate building activities, as compared to the existing situation, and
- are increasingly linked to the urban agglomeration by growing traffic volumes (Bayerische Staatsregierung, 2003).

The LEP's essential messages concerning future development in urban (conurbation) and rural (all other) areas are, on the one hand, recognition of their mutual dependence, and, on the other hand, the need to preserve rural 'autonomy' and regional specificities. In responding to the challenge of containing 'suburbanization', the programme follows a conservationist strategy. This becomes obvious in its statements that the typically rural socio-economic structure of peri-urban areas is threatened, and that land-use planning should aim to contain urban sprawl by concentrating development and applying building structures that are adapted to local contexts (Bayerische Staatsregierung, 2003, p.128).

The processes that have generated these underlying concerns will be explored in the following sections, which highlight the course and intensity of change in the Munich peri-urban belt, using statistical data and our own fieldwork.[3] The focus is on change in (regional and local) settlement structures, on demographic and economic development, infrastructure provision, agricultural change and cultural and natural heritage. To start, the general context of 'suburbanization' in (western) Germany is summarized.

## Current Development Patterns in German Agglomerations

Despite a certain neglect that can be read from the rather small number of publications on the topic in the 1980s and 1990s, 'suburbanization' has continued to be a determining trend in German agglomerations. The late 1990s brought about a renaissance of the topic, which has come alongside the emergence of new tendencies; above all a growing heterogeneity in fringe areas and a vivid discussion of new governance models for urban regions (see, for instance, Brake *et al.* 2001, p.8f). Yet rapid change in *rural* areas beyond the city's suburban 'rings' and older extensions, which have come to be known as peri-urban zones in France, has not received the same attention in Germany as have supposed 'post-suburbanization' tendencies. Thus, the last Regional Development Report took notice of three newly emerging trends that have characterized 'suburbanization' in West German city-regions since the 1990s:

---

3   Quotations in this chapter refer to interviews with regional and local 'change agents' in the Munich region, which were undertaken for the EU funded NEWRUR project.

- 'Expansion of radius: Urbanization is continuously shifting outward toward rural areas. The biggest growth occurs less and less in areas in the immediate neighbourhood of central cities but in the less densely populated and rural areas, which are farther from agglomeration areas.
- Settlement dispersion: Population and employment growth is becoming more and more scattered. It often takes place independently of the spatial planning goal of concentrating growth on focused locations designated by planning processes.
- Functional enrichment of suburbanization: Not only have manufacturing industries discovered the urban fringe as a preferred location. Enterprises in the tertiary sector are also locating there, even though they were assumed to be dependent on factors such as contact, image cultivation and proximity to customers in central cities' (Federal Office for Building and Regional Planning, 2001, p.8f).

Concerning this last point, it should be stressed that the functions mentioned are not equally distributed across the periphery and are not concentrated in the sense of a decentralized concentration. As Burdack and Herfert (1998, p.31) put it, referring to Rohr-Zänker (1996), there is development of '... new monofunctional complexes with large installations for trade, leisure, services, transportation and processing without integration'.

The supposed 'post-suburbanization' of the periphery is linked to these phenomena. This process deals with a change in the current 'suburbanization' phase away from the concentric radial patterns of earlier decades towards new spatial patterns, which are sometimes labelled a 'patchwork structure', sometimes subsumed under the 'Americanization' of the periphery or simply associated with isolation and fragmentation (e.g. Hesse and Schmitz, 1998; Brake *et al.*, 2001). Looking at the growing radius of sprawl, Brake and associates (2001, p.10) question the common definition of post-suburbanization by putting a new emphasis on the syllable 'post': 'The continuous urbanization of the urban fringe is sometimes characterized as "post-suburbanization". But in order to correctly define what really happens [in] "post" suburbanization, we should explicitly refer to the kind of urbanization that is taking place beyond the formerly suburbanized area, in the still rural hinterland'. He draws attention to the fact that, beside the functional diversification of closer suburban areas, another phase of 'suburbanization' can be observed in the more distant hinterland, which is clearly neglected in current discussions, but which is the focus of this chapter.

Aring (1999, p.3) describes changes in the outer fringes of German city-regions in a simplified way: 'During the first half of the 1990s different demands for space overlapped. We could observe a complementation and completion of the former focal points of development *and* a growth at the outer fringes of city-regions. In a simplified way we could as well talk of a suburbanization in second, third or fourth rings'. It seems necessary to stress that this view is simplified as it is obvious that this further 'suburbanization' is very discontinuous. Accordingly, Brake and colleagues (2001, p.18) state that '... the formerly centrifugal and consequently concentric growth pattern of city-regions is being replaced by a patchwork of settlement areas, which differ in their spatial and time-wise evolution'.

A more recent, in-depth study of 'suburbanization' concludes that there is the coexistence of high absolute population growth areas closer to cities, and high relative growth in peripheral areas. This hints at a ubiquitous, spatially diffuse urbanization process (Bundesamt für Bauwesen und Raumordnung, 2003, p.33). In general summary, the same can be said for employment 'suburbanization' (Bundesamt für Bauwesen und Raumordnung, 2003, p.62).

**The Munich Case Study**

The Munich region (Figure 3.1) is comprised of the City of Munich (1.2 million inhabitants) and its hinterland (regional population 2.5 million). Together these are said to be one of the most dynamic regions in Europe (Kagermeier *et al.*, 2001, p.163). This can be confirmed not only from the well above-average purchasing power of its population – and this includes the hinterland as well (as exemplified by Starnberg County and Munich County in Figure 3.2) – but also from having the lowest unemployment rate among German agglomerations (Figure 3.3). In addition to which, the core city of Munich has retained, if not increased, its population throughout the last decades of intense suburbanization in Germany.

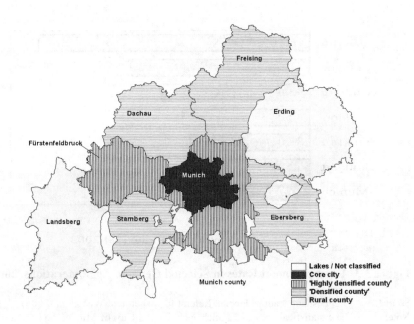

**Figure 3.1 The Munich Region**

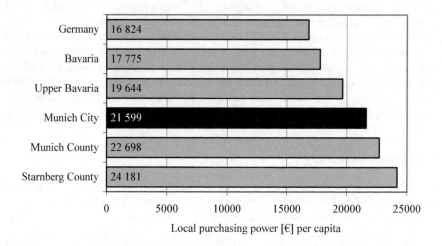

**Figure 3.2 The Purchasing Power in Selected Areas in Germany**

Source: Landeshauptstadt München, Referat für Arbeit und Wirtschaft (2005)

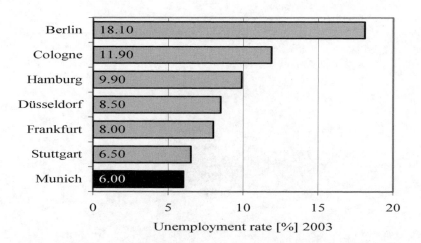

**Figure 3.3 Unemployment Rates in Selected German Agglomerations, 2003**

Source: Landeshauptstadt München, Referat für Arbeit und Wirtschaft (2005)
Note: The statistical area of Munich includes the City of Munich and the adjacent counties of Dachau, Ebersberg, Fürstenfeldbruck, Munich and Starnberg.

The Munich region (5 504 km²) consists of the core city and eight surrounding counties (*Landkreise*), of which the core city and two counties are considered 'highly densified' (Munich and Fürstenfeldbruck), three 'densified' (Dachau, Starnberg, and Freising) and two rural (Erding and Landsberg; Figure 3.1).[4] Despite demographic change, the region has grown constantly during recent decades. At the same time, it has been subject to an intense 'suburbanization' process, which has significantly changed the balance between the City and its hinterland. While in 1970, 80 per cent of the regional population lived in the inner part of the Munich conurbation (more then 60 per cent in the City itself), today this figure is less than 75 per cent for the conurbation and 49 per cent for the City (Table 3.1). At the same time, the City of Munich covers only six per cent of the region's territory, with its population density of 4 019 inh./km² making it one of the most densely populated areas in Germany (Bayerisches Landesamt für Statistik und Datenverarbeitung, 2005).

**Table 3.1  Population Change in the Munich Region 1970-2002**

|  | 1970 | % | 1992 | % | 2002 | % |
|---|---|---|---|---|---|---|
| Conurbation | 1 657 784 | 79.9 | 1 825 360 | 76.5 | 1 857 892 | 74.3 |
| City | 1 293 599 | 62.4 | 1 256 638 | 52.7 | 1 234 692 | 49.4 |
| Outer conurbation | 103 549 | 5.0 | 140 042 | 5.9 | 156 033 | 6.2 |
| Rural near conurbation | 290 940 | 14.0 | 391 989 | 16.4 | 455 062 | 18.2 |
| Other rural | 21 981 | 1.1 | 28 539 | 1.2 | 32 606 | 1.3 |
| Regional total | 2 074 254 | 100.0 | 2 385 930 | 100.0 | 2 501 593 | 100.0 |

Source:   Bayerisches Landesamt für Statistik und Datenverarbeitung (2004)

A closer look reveals that the least densely populated areas have constantly increased their 'weight' in the region, even though inner parts of the conurbation have grown significantly, too. Population growth in rural areas correlates with a growing radius of 'suburbanization' from the core city. In recent years, more distant areas have seen constantly high relative population growth, which has caused significant change in settlement and population structures, given that the majority of rural and peri-urban municipalities have less than 5 000 inhabitants. Meanwhile, the mainly larger municipalities that are closer to Munich still account for the highest absolute growth.

Regional population increases can be attributed to constantly high influxes from outside the region (Kagermeier *et al.*, 2001, p.164f). This is not least because of a supposedly high quality of life. Indeed, a recent study has once again found that Munich is a top-scorer in terms of the contentment of its inhabitants, with 82 per cent stating that they can 'live very well in their city or region'. In this regard,

---

4   This delimitation of the Munich city-region encompasses the planning region of Munich, which covers the relevant regional plan. Other conceptualizations exist, but the planning region is the area of reference in this chapter if not indicated otherwise.

Munich is ranked second in Germany right after Stuttgart. The job market, leisure and cultural activities, and security are among the factors mentioned positively most often (McKinsey *et al.*, 2005). In comparison to other large German conurbations, Munich can draw on a 'modern' economic profile and the absence of old-industrial clusters. Its economic portfolio is based on the 'Munich Mix' of mechanical engineering, vehicle construction, research and technology, media and the like (Kagermeier *et al.*, 2001, p.163; Regionaler Planungsverband München, 2001). A clearly visible drawback to this attractiveness is a chronically strained housing market and above average price levels. Thus, one major driving force for peri-urban change – especially in terms of housing development – is the price of land. The sprawl of Munich into its hinterland has evolved in four distinguishable phases:

- A suburbanization phase in the 1960s characterized by concentration along axes formed by public transport arteries.
- During the 1970s areas between the axes were filled.
- The 1980s were characterized by ubiquitous and disperse growth in the wider periphery. At the same time, the first signs of a qualitative re-urbanization were observed in urban gentrification processes that occurred alongside shrinkage in inhabitant numbers due to high floor-space usage per capita.
- In the 1990s the fourth phase saw a parallel occurrence of re-urbanization tendencies and the emergence of heterogeneous cluster and patchwork patterns, especially in economic development (Kagermeier *et al.*, 2001).

These developments have led to dynamic change in population and employment structures, in the level of services, the supply of infrastructure and, above all, in perceived threats to the environment and heritage of peri-urban areas. The related changes, their causes and effects, are analyzed in the following sections of this chapter. Rather than displaying regional quantitative trends, this chapter highlights how changes are viewed and perceived locally, be these as threats or starting points for a pro-active integration of peri-urban areas into the city-region.

**Regional Settlement Structures**

The region of Munich is characterized by a radial settlement pattern with a strong mono-centric orientation toward the core city. Suburban railway lines extend into the hinterland, well beyond the conurbation. Together with a number of motorways they form the major grid of the region's settlement structure. This is reflected in the Bavarian Regional Development Plan (LEP) (Bayerische Staatsregierung, 2003) and the Regional Plan for Munich (Regionaler Planungsverband München, 2001b), which both aim to strengthen this pattern by trying to channel future growth along these axes and into designated central place locations. In the sense of urban containment, this traditional spatial vision of 'decentralized concentration' is expected not only to increase efficiency in service provision, but also to contain sprawl into supposedly unsuitable locations. Population and employment figures, and the designation of new residential and industrial building areas, clearly show a contradictory trend, for not only have the rural parts of the region attracted the highest growth rates (Figure 3.4), but municipalities without public transport access, which correlate with places with low or no central place functions, have accounted for constantly higher relative population growth than traditional centres (see, for instance, Planungsverband Äußerer Wirtschaftsraum München, 2002).

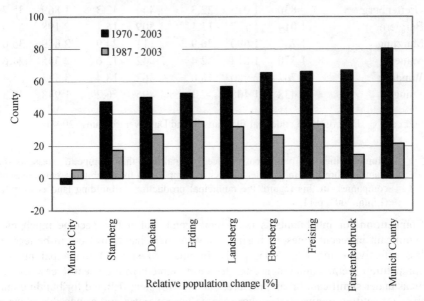

**Figure 3.4  Relative Population Change for Counties in the Munich City-Region**

Source:   Landeshauptstadt München (2004) and Planungsverband Äußerer Wirtschaftsraum München (2005)

Both the above mentioned plans speak a clear language, aimed at restricting both employment and population development to an 'organic amount', which is specified as local demand for new-build plus 'reasonable' additional growth. Yet local growth rates in the last decades seem clearly to contradict this aim. Table 3.2 gives a few examples of extraordinary growth in smaller municipalities that lack central place functions. Today designated new building areas have become almost ubiquitous throughout the region (Kagermeier et al., 2001, p.167f), yet the amount to which these areas are actually built on and the pace at which development takes place differs.

Local policy in this context can be characterized by stiff inter-municipal competition for jobs and inhabitants. This becomes most visible in what could be labelled the local 'production of building sites':

**Table 3.2 Population Change in Some Fast Growing Small Municipalities in the Munich City-Region**

|  | 1970 | 1980 | %<br>1970<br>- 1980 | 1990 | %<br>1980<br>- 1990 | 2000 | %<br>1990<br>- 2000 |
|---|---|---|---|---|---|---|---|
| Greifenberg | 830 | 1 015 | 22.3 | 1 444 | 42.3 | 1 869 | 35.7 |
| Berglern | 1 014 | 1 127 | 11.1 | 1 309 | 16.1 | 2 198 | 67.9 |
| Marzling | 1 235 | 1 690 | 36.9 | 2 111 | 24.9 | 2 849 | 35.0 |
| Mauern | 1 370 | 1 814 | 32.4 | 2 042 | 12.6 | 2 545 | 24.6 |
| Windach | 1 762 | 2 080 | 18.0 | 2 462 | 18.4 | 3 334 | 35.4 |
| Wang | 1 438 | 1 408 | -2.1 | 1 495 | 6.2 | 1 948 | 30.3 |

Source: Bayerisches Landesamt für Statistik und Datenverarbeitung (2004)

> Municipalities are permanently forced to maintain their competitiveness, as they want to score high in the inter-municipal competition for inhabitants, employees and companies. In this regard, the municipal production of building land is essential. (Einig, 2003, p.111)

Competition for inhabitants is not a new trend, but it has become much more significant in recent times, as business development has turned out to be less and less effective (Mönnich, 2005, p.33). In times of strained municipal budgets, competition seems comprehensible, yet short-sighted. In the specific case of rural areas under significant urban pressure, that is, with clear demand for building land, this competition is aimed at higher income tax revenues and at possible planning profits that can be siphoned off by the municipal planning authority.

This competition takes places within a planning framework that grants almost absolute (land-use) planning authority to municipalities, which is executed in the form of the two major instruments at the disposal of municipal authorities; namely, the Preparatory Land-Use Plan (*Flächennutzungsplan* or FNP) and the Binding Land-Use Plan (*Bebauungsplan* or B-Plan). A FNP, which is prepared for an entire municipality, is binding on all public authorities and agencies, but not on private individuals or companies. This plan illustrates, in basic form, all expected (or desired) land-uses. It is essentially a zoning plan that guides the preparation of binding land-use plans. The FNP must adhere to mandatory objectives in a regional plan, must be coordinated with neighbouring municipalities and is subject to an intensive participation process by public bodies and individuals (Commission of the European Communities, 1999a, p.64). FNPs document a municipality's mid-range development aspirations by outlining the major spatial development objective for the next 10 to 15 years. These plans are based on more or less comprehensive population and employment projections and have to balance local 'ambitions' with restrictive or pro-active provisions made by superordinate plans.

Binding land-use plans provide the basis for the detailed and legally binding control of building development. They can be applied to unbuilt land to open it up for first-time development, or equally they can cover areas already developed or to be redeveloped.[5] They are adopted by municipal councils as local by-laws and must only be approved by higher state authorities (usually counties) if the plan has not been developed in accordance with the relevant FNP. Their provisions are binding on all public authorities and private individuals. As a B-Plan is the legal basis for issuing building permissions, it must indicate type and extent of land-use, land area to be covered with buildings, and areas required for traffic purposes. Optionally, B-Plans can include elements like minimum dimensions for building plots, alignments, the maximum number of dwellings in residential buildings, and space for public thoroughfares, including pedestrian areas and car parking (Commission of the European Communities, 1999a, p.65).

The essence of urban land-use planning as described above is the production of sites. This adds value to agricultural land by assigning it for more profitable uses, securing accessibility, providing protection against conflicting uses and guaranteeing land-use security for investment. Although the right to own a plot of land and the right to use it for a specific purpose are strictly separated by law, profits accruing from assigning higher value uses are expected to be reaped primarily by the owner, not by the public authority granting this privilege. In many countries higher land values result in higher property tax revenues, but this rarely reflects gains to the private owner. In Germany, the returns from property tax collection are low, if not insignificant, but municipalities have found other ways to participate in the land value increases they generate through land-use planning. Since the local level in the administrative hierarchy has the right to assign land-uses, it is interesting to see how municipalities use this double privilege. It is also worth noting that part of German income tax revenue is allocated to municipalities on the basis of their population, as are payments from other financial equalization instruments. 'Producing' sites and attracting new inhabitants is thus a twofold incentive for municipalities, which has helped fuel 'suburbanization', not only in the Munich region.

With their zoning monopoly, municipalities have full control over urban growth within the framework set by a regional planning authority. Peri-urban municipalities in the Munich region have established the practice of acquiring future building land, developing it and then selling it to building owners. Prices are determined for the time municipalities buy plots and through negotiation with a former owner. It follows that an anticipatory land acquisition of potential building land well ahead of preparatory and legally binding land-use plans would, in principle, allow municipalities to buy at very low prices (e.g. as agricultural land). But this is not common practice. Instead, acquisition tends to take place shortly before the development of plots. As municipalities have to pre-finance all investment related to the opening-up of building land, pressure arises to develop areas as fast as possible. Questionable settlement layouts are just one consequence.

---

5   The latter usually involves a long planning procedure, which can be slowed by owners who fear losing existing building rights under Section 34 of the Federal Building Code. Thus, municipalities often refrain from using this instrument in already developed areas.

> This is very difficult for the municipality at the moment. We always have to pre-finance until the money flows back. ... Pre-financing the sewers and sewage treatment plant is asking too much of the municipality. We have reached the limit. (Municipal Mayor)
>
> Quality [of planning] doesn't count. I just have to be quick, because otherwise my neighbour does it. (Municipal Mayor)

The range for negotiations between a municipality and a landowner, who is usually a farmer, can be increased through an extensive designation of potential building land in a preparatory land-use plan. This strategy provides the municipality with the possibility of moving on to other landowners if it fails to acquire a specific plot. This practice undermines the overriding principle of the restrictive designation of building land. The Regional Planning Authority, which has to approve the preparatory land-use plan – but not subsequent alterations or extensions – is caught in a dilemma when a municipality argues the need to develop additional building land at affordable prices. A 'typical' municipal argument follows the logic:

> The acquisition of building land causes no serious difficulties. The problem that owners don't want to sell is avoided by designating a little more than currently needed in the preparatory land-use plan. This strategy creates alternatives. (Municipal Mayor)

Yet the range of communal interventions in the land market tends to shrink with an increasing scarcity of land suitable or available for new-build designation. As a consequence, the role of municipalities as agents in the market for building land tends to increase with distance from the centre of an urban agglomeration. While the City of Munich finds it very difficult to enter the land market, and even suburban municipalities are facing limited room for negotiations with landowners, most peri-urban municipalities are lead players in their local land markets. The following statements illustrate local views on the situation for municipalities with limited land reserves:

> A restraint is that the municipality itself cannot get any plots and does not have any at its disposal. The farmers don't want to sell their plots. The land designated in the preparatory land-use plan for development is very hard to get or very hard to pay for. (Municipal Mayor)
>
> It is financially not viable. The farmer doesn't sell any land to the municipality if he can assume that it is turned into building land. (Municipal Mayor)

The practice of buying building land before plan establishment usually generates benefits for a municipality, even at relatively high 'acquisition prices', because it brings established plots onto the market at a profit. But with limited reserves of developable land, a municipality is forced to grant more concessions to landowners. Hence, municipalities often obtain plots only by offering owners a share of net building land, usually 25 to 50 per cent.

Acquiring and reselling building land leaves municipalities with 'planning profits', which help pay for infrastructure and services for the growing and changing population. Siphoning off planning profits is not an end in itself, but also subsidizes building schemes for local residents,[6] who are increasingly threatened with being ruled out of local land markets by rising prices.

Several issues arise which can be viewed negatively in the context of the spread of 'urban growth' (or 'suburbanization') within the city-region. As mentioned before, residential development has become an almost ubiquitous phenomenon, that is fuelled by locally justifiable yet unsustainable and counterproductive considerations from a regional point of view. The dispersed nature of the resulting residential expansion questions the agreed regional objective of decentralized concentration and opens up areas for development that are meant to be restricted mainly to locally-induced growth. At the same time, the practice of ubiquitous and (partly) inconsiderate designation of building land provokes visible change in local settlement structures, which will be elaborated on in the next section.

## Local Settlement Structures

Bavaria is a German *Länder* that, despite 1970s territorial reforms, has retained many very small municipalities. Each municipality disposes the same planning authority, which aggravates competition. The municipal multitude reflects a dispersed rural settlement structure that persists to this day, while large extensions to existing built-up areas result in significant change in local settlement structures. In this context, two processes are evolving in parallel.

In principle, German municipalities can develop their urban morphology within the general framework of regional plans and planning law. But as agriculture is restructured, many abandoned farmsteads are appearing in villages and hamlets. While many of these buildings remain empty, many others have been sold to developers, sometimes to finance resettlement to a farm outside the village. Under Section 34 of the Federal Building Code (Baugesetzbuch or BauGB)[7] building within a continuously built-up area is acceptable, even where there is no binding land-use plan. Thus, open plots within villages have substantial potential for new residential and, to a more limited degree, commercial use. But this building right is

---

6 Municipalities can supply subsidized land to local residents. To qualify, applicants usually have to have lived in a municipality for a specified time and agree not to resell the land within a set period. The strictness of the criteria varies across municipalities.
7 Section 34, on the permissibility of development in built-up areas, states that: '(1) Within built-up areas a development project is only permissible where, in terms of the type and scale of use for building, the coverage type and the plot area to be built on, the building proposal blends with the characteristic features of its immediate environment and the provision of local public infrastructure has been secured. ... and (2) Where the characteristic features of the immediate environment correspond to one of the specific land-use areas contained in the legal ordinance issued in pursuance of Section 2 para. 5, the permissibility of the development project is determined solely with reference to type and to whether it would in general be permissible under the ordinance within the specific land-use area'. (Federal Building Code – *Baugesetzbuch* BauGB, in the version amended by the Act to Amend the Federal Building Code and to Reorder Spatial Planning Law [BauROG], issued on August 18th 1997 (BGBl. I p. 2081).)

perceived ambiguously locally, in some cases even being perceived as a 'threat', because old farms are large enough to accommodate a significant number of new dwellings. Moreover, the quantity that can be constructed on these sites is beyond the direct control of a municipality. Indeed, as is not the owner, and commonly lacks the resources to acquire these valuable plots, the municipality also does not have the capacity to siphon-off planning profits from such developments:

> This was also the reason why we slowed it down. You create flats, 50 or 100 flats, and you don't have a chance to make a profit from it. (Urban Planner)

> But in the village, at the roundabout, we have a new settlement. It was an old farm, and there are now 20 new dwellings. (Municipal Mayor)

> If you look at the old farms, you can easily build 10, 12, 16 flats there. (Urban Planner)

This 'densification' process in village centres evolves in parallel with the designation of new building areas, although 'densification' of this kind is less visible and is thus often interpreted as a 'creeping' process. Despite a clearly articulated fear that intense reuse might lead to a loss of rural character, there are expectations that such developments might help meet growing demand for rental housing in an acceptable way. This is true in municipalities with good transport links that have a high share of out-commuters. Farmstead reuse is also a means of containing the 'consumption of land' which accompanies first-time development of farmland.

> This is done by real estate agents and developers, and we have to take care that building does not become too dense. We want to preserve the rural character, don't want an urban structure. ... In a new building area you can control everything. ... but where you have existing building rights on an old farm, where you build five houses in a row, the character can soon be distorted. (Municipal Mayor).

At the same time as 'densification'; occurred, new-build is emerging in designated areas. Characteristically, the type of buildings erected here is the same, irrespective of the size of the peri-urban municipality. The preferred style is single-family homes, either detached or semi-detached. Indeed, a major driving force for building in peri-urban areas is the desire to live in a single-family home – detached if possible. As some of our peri-urban informants put it: 'You have to be able to run around your own house', and 'People want their own plot with their own backyard'. But to be able to realize this, different compromises must be accepted. First, a decrease in the size of building plots is occurring. Today, detached single-family homes are built on plots of 500m$^2$ - 700m$^2$ plots, with semi-detached houses on about 400m$^2$. Plots in local building schemes tend to be smaller, and thus cheaper, in order to achieve the desired intention of providing homes for less wealthy families: 'Better a small backyard but a neat house', as one municipal Mayor put it. Alongside the more intensive use of plots, there is a growing demand for semi-detached houses, which are cheaper than detached ones, although larger dwelling units are generally considered by municipal leaders to be inappropriate for peri-urban municipalities.

In this respect the type of building seems more important than its actual size or the number of flats within. Two-family houses are widely accepted, which can be seen from available statistics on the housing stock, and from the opinions of different stakeholder groups. Yet terraced houses are considered too dense and 'urban' in style. They are widely rejected as an inappropriate housing style.

> We just managed to prevent the building of terraced houses. (Municipal Mayor)

> Living in a village means loose building. Nobody moves out here to live in a block of flats. (Municipal Mayor)

> Only semi-detached houses are accepted. There will surely be no trend to build terraced houses. It also doesn't fit our planning, would create social problems, and a totally new type of people would move in. (Municipal Mayor)

While detached and semi-detached buildings are considered appropriate, planners hardly make further provisions on form. A few large residential areas have been erected by developers, but the more common pattern is for a gradual build-up of residential units by individual households. This often leads to low design standards and to an uncoordinated accumulation of housing styles. In general, the quality of design and the form of legally binding land-use plans can be considered questionable. Unconventional, futuristic building designs and neighbourhood layouts are rare. Most municipalities do not exhaust their 'planning power' for this purpose. Indeed, with residential areas developed incrementally, the scope for a single land-use plan for new residential development is limited. But in the absence of a medium-term to long-term development 'vision', municipalities expand piecemeal. It is on demand that land-use plans are made, with land developed where available or where an owner is willing to sell. Incremental growth leads to 'creeping' land-use and housing change in peri-urban areas.

Structural change in agriculture and high demand for peri-urban housing lead to a high potential for the reuse of abandoned farmsteads, so opening the potential for sustainable development strategies by using 'brownfield' sites to prevent 'greenfield' development. But the very character of peri-urban areas prevents this potential from being realized satisfactorily. With jobs increasingly spread over suburban rings, and with land prices and rents beyond affordable levels for many, the peri-urban zone is considered the main alternative for building one's own house. Here land is still comparably cheap, traffic and travel times are still bearable and few restrictions are made in land-use plans on styles of housing, provided units are detached single-family homes or as close to this as possible. In contrast to suburban areas, the sparing use of land is neither demanded nor offered.

## Demography and Infrastructure

Despite recent studies showing that 'suburbanites' in western Germany have become more heterogeneous (more singles, more couples without children and more from other groups that differ from the stereotypical 'suburban young family with children'; Heitkamp 2002; Ismaier 2002), population data for the Munich region shows that peri-urban populations are still significantly younger than the conurbation's average. Age groups representing the suburban stereotype family (0-14 and 30-49) are clearly over-represented, whereas older groups are under-represented (Figure 3.5).

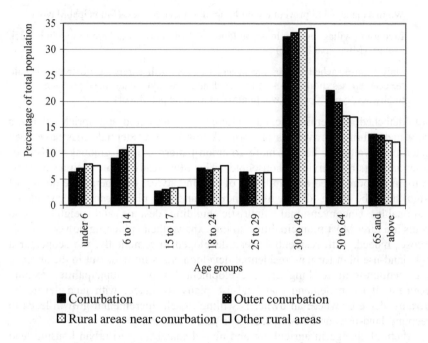

**Figure 3.5 Percentage of Population by Age Group for Different Areas in the Munich Region**

Source: Bayerisches Landesamt für Statistik und Datenverarbeitung (2004)

These figures match what local government representatives see as the target group for newly designated building areas:

> We have become a very young community, because the people moving into the new areas are very young families with a lot of kindergarten children. People that work and have a good income. They come from the agglomerations. (Municipal Mayor)

Yet proximity to the urban agglomeration and its job market still dominates the region, making peri-urban areas different from remoter rural parts of the region. Although in-migrants tend to be younger than the resident population, couples with children are not the only group that is moving in. Above all, local commentators believe that couples without children and single parents are important newcomers.

> You can feel the changes in society. Mothers work, because the family needs two incomes. (Municipal Planner)
>
> Our offers have become more flexible, we aim at single parents too. (Municipal Mayor)
>
> One problem is that children who move here with their parents are already 10 or 13 years old. They are already too old to participate in the existing childcare offers. (Official in the Bavarian Office for the Environment)

Rapid population growth and resulting changes in population structure have caused significant investment in technical and social infrastructure in almost all peri-urban municipalities. These include those that become necessary because of population *growth* and those that can be attributed to changes in population *structure*.

Without exception, kindergartens are mentioned as major investments, either as new operations or as significantly extended facilities. Peri-urban in-migrants no longer fit traditional images of 'typical suburbanites', for they generate demands ranging from all-day kindergarten groups, where both parents work, to day-homes for school children in the afternoon. While these new demands are caused by new lifestyles in peri-urban and rural areas, they also reflect a new attitude toward service supply, which asks for more equality with the city.

> There is a new demand, with a range including cemeteries. In former times, people went walking, cycling and swimming. Today they ask for much more from the municipality. (Municipal Planner)
>
> Surely, the pressure on the municipality has grown. We have attracted a new class of people with a much higher demand than we have ourselves. I don't want to spare the locals, but the urban people take everything for granted. (Municipal Mayor)
>
> Ten or 13 years ago, everybody still knew everybody. This is no longer the case. Urban habits are transferred to the countryside. 'My home is my castle' and 'everything around doesn't count'. There is little participation in community life. We just hope that the children grow up together. (Municipal Mayor)

In the school sector, the situation is similar. In addition, sports facilities, mainly football grounds, tennis courts, but also sports halls, have been extended or newly built in almost all municipalities, partly to a very high standard.

These investments are problematical for small peri-urban municipalities for several reasons. The first is embedded in the often-practised siphoning-off of planning profits from land sales. This is not done as an end in itself, for profits are often reinvested in social infrastructure made necessary by a growing or changing population. Where investment needs exceed short-term profits, municipalities can be caught in a vicious circle. All these investments have to be borne to a large but varying degree by municipalities themselves, using different sources of funding. Previously, when business tax revenues were a constant source of income, these covered the bulk of growth-generated demand. Today, besides 'being economical' or 'running up debts', the best way to finance such investments seems to be to develop and sell building land. Without this practice many municipalities claim they would no longer be able to meet new demand. But by acting in this way, they are running a high risk of becoming more dependent on this 'growth machine', with growth generating demand for infrastructure, which is met by land sales, which by themselves create further growth.

> Only because we designated such large residential areas could we pay for all this. There are no developers locally. The municipality is always the developer. Because of this we get all the income, but we don't keep it but reinvest it immediately. Building land is bought at €35, opening it up costs about the same. The net building land is then sold at €140. We siphon off the profit ... These are nice incomes. We have to pay all the infrastructure from this. (Municipal Mayor)

Rather as a consequence of growth than of changed population structure is the need to invest in 'hard' technical infrastructure, like fire brigades, sewers, cemeteries, etc., of which dimensions for provision are calculated according to the number of inhabitants. Sewers and waste water treatment plants are critical investments in the sense that they need a certain number of inhabitants to operate them efficiently in the long run. This can create additional ambitions amongst local politicians to attract further new inhabitants, so entering another vicious cycle.

Another problem in terms of financing infrastructure lies in the provision of line-infrastructure like roads and sewers. Ninety per cent of the cost of these facilities is borne by plot-owners or house-builders.[8] Whenever new residential areas are filling fast, these costs can be recovered quickly. But if plans cannot be implemented as envisaged, municipalities have to bear the cost themselves. This is another factor 'seducing' municipalities into keeping standards and provisions in binding land-use plans as low as possible, so as not to deter people from building in the municipality, given the degree of inter-municipal competition for new inhabitants.

## The Peri-Urban Economy

The development of industrial areas is only marginally linked to that of residential areas and is influenced by different incentives and parameters. In fact, the correlation in 1993-1997 annual growth rates for residential and employment change in the region has been calculated to be as low as 0.02. Between 1970 and 1987 it was as high as 0.88 (Kagermeier *et al.*, 2001, p.168). This underlines the emergence of a 'heterogeneous patchwork structure', as mentioned earlier.

Yet statistical data on employment change in the Munich region is blurred by the effects of the new airport in the north-eastern part of the region. This was opened near the City of Freising in 1992 and has created an estimated 100 000 new jobs. Beyond the immediate airport catchment area, little employment-related development is visible. Even so, building land has been designated extensively, although the actual degree differs significantly. In the 'airport region' alone (i.e.

---

8 'The *Gemeinde* has a right and duty to impose a charge on landowners to recover the costs of local infrastructure. The landowners pay a maximum of 90 per cent and the *Gemeinde* pay a minimum of 10 per cent. ... The detailed criteria for the assessment and allocation of local infrastructure charges ... are provided in Sections 127-135 of the BauGB' (Commission of the European Communities, 1999a, p.97).

Freising and Erding counties, plus parts of Munich and Landshut counties), there is an estimated surplus of land designated for employment of 300 per cent, given projected demand is only for 180 hectares by 2015 (Bayerisches Staatsministerium für Wirtschaft, Verkehr und Technologie, 2002).

In this regard we can distinguish between two types of municipality: those with good and those with poorer accessibility, or those closer to Munich and the airport area or those that are farther away. The latter especially have extensive zones for new employment, but beyond the need for investment by local companies these are hardly marketable. Beside obvious traffic bottlenecks, like missing motorway access points, railway crossings, or insufficient secondary roads, the general agreement is that there is a regional surplus in zoned industrial areas, which are less attractive in terms of accessibility and location relative to the conurbation.

Yet such development continues to take place, even though unemployment rates are much lower than the national average. Freising County, for instance, had an unemployment rate of only 4.5 per cent in September 2004, with Munich at 7.4 per cent and Germany at 10.3 per cent (Bayerisches Landesamt für Statistik und Datenverarbeitung 2005, Statistisches Bundesamt, 2005). It is worth noting here that municipal arguments in favour of designating employment zones have changed in recent years. Prior to 2001/2, hoped-for business taxes were a key argument for designating areas, differing regional plan objectives and suboptimal accessibility notwithstanding. But with a sharp drop of business tax revenues in the years 2001 and 2002,[9] and the difficulties they create owing to annual fluctuations (e.g. Bayerischer Städtetag, 2003), this argument is now less viable. Instead, local politicians now argue that there is a need for additional jobs, even though unemployment is generally low (albeit on the rise).

> We have tried to create a second pillar for the municipal budget on business taxes. But its revenues have gone down again. The income for the municipality is lower, but the jobs for the locals are also important. (Municipal Mayor)

> We hope for the best. The prime reason for the allocation was and will be the jobs. (Municipal Mayor)

As a consequence of local misconceptions about the marketability of industrial estates many of small peri-urban municipalities have become or rather remain dormitory settlements. Even in those with a larger number of jobs, the rising distance between home and workplace is perceived as a problem. The growing spatial separation of home and workplace can be attributed to a greater mobility of people homes rather than to their places of employment. People who move out of the urban agglomeration but keep their jobs there tend to look for a place to live where they can optimize the balance between commuting distance and either the price of building land or rents. Thus, commuting becomes inevitable if a job is retained. Beside some out-commuters, municipalities with larger employment opportunities tend to have a large number of in-commuters. Even here, places of

---

9   It was changes in the law, among them the possibility of charging business profits and deficits (including for subsidiary companies), plus tax exemptions for the transfer of an enterprise, that led to a dramatic decrease in business tax revenues in 2001 and 2002.

residence and workplace rarely coincide. One can say that the formerly valid image of 'working in the core city and living in suburbia', which was long true for southern Bavaria, has been replaced by 'living and working in different parts of the city-region' (Kagermeier *et al.*, 2001, p.169).

**Peri-Urban Agriculture**

The type and extent of structural change in agriculture that is taking place in the Munich region has not been caused by peri-urbanization but by exogenous factors, such as EU market and price policies. Some processes are nonetheless evolving at an accelerated speed within the region, at the same time as specific circumstances help absorb some impacts. Whereas proximity to the conurbation offers alternative income and job opportunities, urban pressure on agricultural areas causes land-use conflicts, which are greater than in less dynamic areas.

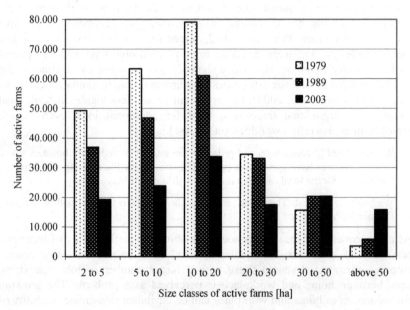

**Figure 3.6 Change in Farm Sizes in Bavaria, 1979-2003**

Source:   Bayerisches Staatsministerium für Landwirtschaft und Forsten (2004)

Generally, the number of active farms is decreasing. This is especially true for full-time farmers, who are mainly cattle keepers, as the future of this type of farming is considered particularly difficult, due to its labour intensity and unsuitability for part-time farming.

> ... the majority of the give-ups are dairy farms. Because a dairy farm is very labour intensive. (Farmers' Union Officer)
>
> Dairy farming is a full-time commitment, otherwise you cannot live from it, but that means twice a day 365 days a year going in the stable. (Rural Development Officer)

For crop farming, the trend lies in the increased size of remaining businesses. High quality soils are becoming rarer, and the price for arable land is high, especially in high pressure areas close to the airport. Here, an indirect influence of urban pressure on agriculture can be found, as expansion in farm size is constrained by the tight land market.

Parallel with farm abandonment and changes in part-time farming, there has been a general increase in the average size of farms (Figure 3.6). Both the number of very small farming units (largely farmed part-time) and those with more than 50 hectares have increased. Usually, if a farm is given up, land on leasehold is handed over to another farm, which is extended in order to continue in operation. In this regard, a significant difference between peri-urban areas close to urban centres and remoter rural areas (e.g. in Lower Bavaria) lies in the availability of jobs in the former. This can be seen as another factor accelerating structural change. At the same time, social impacts of change are cushioned by the reasonable job market in Upper Bavaria. As one Mayor put it: 'Our farmers virtually line up along the streets and the next employer takes them with him'. If there are opportunities to find off-farm employment, the decision on continuing or giving up farming is said to be rather easy. Even where job opportunities are supposedly not as ample as in the dynamic airport area, the situation does not differ significantly. The Munich conurbation and a variety of secondary towns offer alternatives.

Yet it is not common to give up farming because of the opportunity to sell land for building. This is despite the fact that the conversion of farmland for building purposes is a critical issue and the willingness of farmers to sell is high. Cases where farmers refuse to sell to a municipality are reported to be exceptions. If it is foreseeable for a farmer that a municipality wants to buy her or his land for building, the price is significantly higher than for pure farming land, thus giving the farmer a strong position in negotiations with the municipality. Most commonly though, farmers sell if gains can be reinvested in larger or better land parcels that expand their operations. Resettlement due to a lack of expansion possibilities inside villages and hamlets has become a noteworthy factor in peri-urban areas. To finance this, old farmsteads are sold, with realized funds invested in new farm buildings and land. High costs provide incentives to sell to developers, while placing pressure on developers to secure high returns from residential units, so causing a remarkable change to the inner structure of villages. If a municipality uses legally binding land-use plans to reduce a farmer's building rights, this threatens a farmer's ability to resettle, as such restrictions tend to limit land sale returns. Hence, many municipalities refrain from preparing land-use plans, as they interfere with building rights and provoke protests from landowners.

> If a farmer wants to survive nowadays, he has to resettle. But this is expensive and only possible if the sale of the farmstead brings enough money. The resettlement costs about DM2-2.5 million. When the municipalities cut the building right on the

farmsteads, there won't be enough money for the farmer. (Rural Development Officer)

Yet, if farmers stay within the built-up area of a village, conflicts can arise with new neighbouring land-uses. As long as agriculture was a 'typical' land-use, village centres were designated as 'mixed-use' or 'village areas'. In these cases, standards for noise emissions and minimum distances to adjoining built-up areas were lower. With the retreat of agriculture and residential growth, such areas are now depicted as 'general residential' or even 'residential-only areas'. If farms are in close proximity to dwellings in such areas their operations are subject to restrictions.[10] If distance from dwellings is to be maintained, expansion of a farm business becomes almost impossible. Moreover, farming within the built-up area is increasingly regarded as a disturbance, especially by new residents.

> If today a building area without a local building scheme is designated, then the tenants or owners are strangers, newcomers as we say. They cause great trouble. Emissions from farms, annoyance by noise, smell, and so on, things like that I get on my desk almost every summer - complaints from non-farmers about agriculture. Because it is smelly, because it is dirty, because the cows are howling, those things I have to deal with almost all the time. And usually by people who have moved in. Not from locals, but those who moved in. (Farmers' Union Officer)

Proximity to the City and the moving-in of often wealthier people give the chance of continued farm development as the market is open to alternative income generation, which is less prominent in rural areas that are more peripheral to cities. The direct marketing of agricultural products is regarded as a further economic niche, which is encouraged by ecologically sensitive, wealthy in-migrant customers:

> In Freising we have a clientele; they are people who deal with ecological subjects – students and newcomers, people with a certain income and grade of education. Additionally people are being made sensitive by the airport. (Agenda 21 Official)

The same is true for horse keeping, which is also prompted by proximity to the urban agglomeration. But this form of land-use is another example of how new, typically peri-urban land-uses generate a high potential for conflict.

> There are areas, which are not being farmed, but are leased to horse keepers. There are riding horse farms. They need hay and straw. They lease land and put their horses there, to the distress of the county and the Farmers' Union. Nobody has something against it, as long as they stay on their own ground, but they ride over foreign land, and cause damage there. ... There is one lawsuit a half-year. Therefore we don't like the horses. If you look at a horse farm, a bit south from here, it has 80 horses. He has 14 of his own and the rest are horses from outside. They come from the greater region of Munich. They put their horses there, pay the rent, pay for the food and come twice a week with the daughter and ride. (Farmers' Union Officer)

---

10 General and specific land-use areas are regulated in detail by the Federal Building Ordinance (*Baunutzungsverordnung* – BauNVO).

## The Peri-Urban Environment

Peri-urban growth has led to undesired impacts on the environment. Increased commuting and other traffic flows are among the most visible and measurable impacts. Another effect is the accelerated 'consumption of land' or the conversion of agricultural land to residential use. The state of Bavaria is currently the 'leader' in the consumption of land in western Germany (Bundesamt für Bauwesen und Raumordnung, 2004). In quantitative terms the daily growth in areas for settlement and transportation represents 28.4 hectares (or 103 km² annually). This is roughly equivalent to the total area of the City of Nuremberg, with its almost 500 000 inhabitants (Bayerisches Staatsministerium für Landesentwicklung und Umweltfragen, 2003). As a consequence, a reduction of land consumption by different means was made one of the major objectives of the 2003 Bavarian Regional Development Programme, as this consumption is primarily caused by frictions and misconceptions in and about local land-use planning.

Land is 'consumed' in a twofold way:

- On a regional scale, superordinate planning objectives of concentrating growth in central places locations and along major development axes is only successful to a limited degree. Non-central places and rural areas meant to preserve 'rural characteristics' have grown well beyond the desired organic rate. Building land has been offered almost ubiquitously among the different parts of the region. In this regard, settlement and transportation areas have almost always grown faster than the number of inhabitants and jobs, thus hinting at an increasingly sparse and loose development.
- At the local level, it is the type and quality of land-use plans that wastes potential for a more efficient use of land. A low density requires higher costs for infrastructure provision. At the same time, the detailed layout of plans often lacks comprehension of resource-efficiency. A typical layout plan in the north-eastern part of the region would thus put the building in the middle of the plot; double-access and thus a excess of thoroughfares is common.

Both can be traced back to a low valuation of land as an increasingly scarce resource by local planners, politicians and surely by house-builders themselves. Whereas the 'quality' of clean air or water and their need to be protected has long been accepted, a protection of land and soil as an end in itself is not yet commonly understood.

> Land as such doesn't have any value ... the usual protection categories are being accepted, but land or landscape is difficult to protect with arguments. (Municipal Mayor)
>
> The problem is that we need water and air to live, that is known for long. With land, it is apparently still not the fact. That is only dirt at our feet. (Regional Planner)

> Where the limits of growth can be experienced, a certain awareness building comes along. Where the limits of growth cannot be experienced, it can neither be mediated. There is the case of Augsburg, flat countryside, no limits by the topography, space without limits for the low-brow observer. (Landscape Planner)

The fact that land consumption is highest where few restrictions are felt – that is, in rural areas – underlines this assumption. Even within peri-urban areas a distinction in the apprehension of this topic can be seen between municipalities with seemingly unlimited land reserves and those that have already felt its scarcity for different reasons, be it because of nature protection areas, because of rising prices or for instance because of an inability to obtain land from farmers.

## Conclusion

ESDP aims with regard to a 'sustainable development of towns and cities'[11] – though referring to less dynamic areas – are surely all the more important to consider in a dynamic region, where a self-confident hinterland is a key player in the development of the whole region. This self-confidence becomes very obvious in the peri-urban 'self-conception', as expressed by various key actors.

> Yes, we are still a rural community, but with a very good infrastructure. This year, we have opened a double multipurpose hall, two years ago a new school, four kindergarten groups, doctor, veterinary, dentist, pharmacy, a supermarket, three pubs, which are well known in the whole region, and a bakery. But still we are rural. There is an active social life, with a skiing club, football, tennis and still some agriculture. The hamlets are still agricultural. In the centre you can build denser, the hamlets should remain rural. (Municipal Mayor)

> Change has not been that drastic in the last ten years, it had started before, away from the rural village character, even though my predecessor followed conservation as an ambitious goal, but it was only possible in a limited way. A change has taken place. Twenty, 25 years ago, we were still a village, but the motorway has always been there. (Municipal Mayor)

> We are a rural community, with a structurally shrinking agriculture and a focus on housing in the future. ... We have many commuters to Munich. They also settle here. They want to live in a green environment. For financial reasons they come out here, ... we are the nest village with acceptable prices. We want to remain a high-quality neighbourhood with good infrastructure. (Municipal Mayor)

The Bavarian Regional Development Programme (LEP) calls for a preservation of the supposedly rural characteristics of peri-urban areas in order not to sacrifice it to extensive suburbanization and sprawl. This conserving attitude – although a mutual approximation of rural and urban areas, for a balancing of costs and

---

11 Control of the physical expansion of towns and cities; a mixture of functions and social groups; wise and resource-saving management of the urban ecosystem; better accessibility by different types of transport which are not only effective but also environmentally friendly; and the conservation and development of the natural and cultural heritage (Commission of the European Communities, 1999, p.22).

burdens is also aimed at – seems almost paradoxical, as it is the particular qualities of rural areas in conjunction with proximity to a conurbation that make peri-urban zones so prone to further 'suburbanization'. This particular quality is often seen to be exemplified in an intact environment, greater security, more privacy and better recreation potentials. These perceived attributes can be read from the motivation of 'suburbanites':

> Among the reasons for leaving Munich mentioned most often (without a specific order) are the desire for more 'green', better air and less noise – that is better environmental conditions. Every second household mentions this as a reason. ... Roughly one third of the interviewees misses a garden of their own and wants a better neighbourhood. Just as many mention the size or high costs of their Munich dwelling as decisive. ... In about every fifth case parents want to offer their children better living conditions. (Landeshauptstadt München, 2002, p.12)

The peri-urban area acts in many cases as a mirror image of the conurbation. In the particular case of Munich, high land prices and rents just act as a catalyst. The hinterland thus attracts, without obliging migrants to give up the city's amenities:

> The hinterland with its outward oriented housing, the proximity of the open landscape and its low environmental constraints on the credit side and with strong social control, increased travel costs and rural boredom on the debit side, just becomes to convenient because of the proximity of the city. (Krau, 2005, p.49)

This perception, which was surely true for former phases of 'suburbanization', is all the more valid today. But due to the advanced urbanization of the fringe, and its functional enrichment, urban amenities have come even closer to the area that is still supposedly rural – at least in terms of settlement structures. It is exactly this settlement structure that makes a difference between the closer and the further hinterland and that displays the paradox that lies in the LEP objective of containing 'suburbanization' by conserving rural characteristics in the peri-urban area.

The peri-urban hinterland is a stronghold of detached single-family houses, a few, larger municipalities with denser development notwithstanding. Lower prices in peri-urban areas are a crucial factor that accelerates 'suburbanization'. Yet was not the detached single-family house believed to be *the* stereotypically suburban? Exactly this type of building is now considered the only acceptable, and supposedly the typical, housing style for rural, now peri-urban, areas, which is now seen to be essential to preserve. Unfortunately, this preservation is not aimed at a qualitative advancement of the housing stock, which bears a larger potential for a denser, more environment friendly and 'typical' development pattern, that could still comply with ESDP policy aims. Current housing quality advancement seem to be restricted to the modest requirements of new land-use plans, where semi-detached houses determine the maximum density.

In order to appreciate the absolute damage done by the ongoing expansive policy of a multitude of municipalities, it is necessary to recognize that local actors lack an integrative view of the city-region, as called for in the ESDP. This helps us comprehend why the designation of new, faceless, but supposedly typical building areas, leads to a convergence of peri-urban and suburban settlement structures, that make the call for a mutual sharing of costs and burdens obsolete. The specific amenities of the hinterland, which have made it so attractive to date – even if for horse riding – will disintegrate sooner or later. Bottlenecks in childcare are being imported from the core city and problems of social segregation will not take long to be sharply drawn, if locals are driven out of the land market into less attractive areas by a constant demand from outside. Also, many small municipalities are hardly, or not at all, able to meet growing but often legitimate demands from new inhabitants. Moreover, with a shrinking and ageing group of 'indigenous' locals, the local community's often cited integration capacity decreases.

These developments happen against a backcloth in which spatial planning policies – including the ESDP – only rarely exert influence. Even an increased supply of single family houses in the core city has limited success in lessening peri-urban tendencies, as reasons for leaving the city seem complex. It is not just the form of the house that is critical, but a complete package of motivations. At one level municipalities supposedly act rationally, as promoting growth seems to promise returns, even if it does not necessarily comply responsibilities for planning for the public interest. If the public interest is viewed as promoting sustainable development, then this could be applied through an urban containment strategy. This in turn would commit municipalities to better consideration of guidelines laid down in the ESDP and the LEP, particularly over the prevention of further 'physical expansion of towns and cities'. This is what some municipal policies in peri-urban areas are already beginning to push forward, while at the same time denying a further approximation of urban and rural structures:

> We are a rural community, still, although we are so close to the conurbation. ... Growth, yes, we want it as building development, but not as expansion. (Municipal Mayor)

> Yes, we started with 1 400, now we are at 2 000 and this will level off at some 2 500 or 2 700 ... but investments must pay off. (Municipal Mayor)

Keeping this in mind, it is necessary to create a financial and planning framework that alleviates further competition. The ESDP 'partnership between town and countryside' thus has to be advanced through a discussion about common and binding instruments than are available today. This would include municipal tax and budget reform, joint inter-municipal building areas, especially for industrial estates or tradable rights for building land designation (see, for instance, Einig, 2003; Bundesamt für Bauwesen und Raumordnung, 2005). Above all, there is need to create more awareness of the limitations of local potentialities and attractions – a prerequisite which cannot yet be taken for granted.

## References

Aring, J. (1999) *Suburbia – Postsuburbia – Zwischenstadt – Die jüngere Wohnsiedlungsentwicklung im Umland der großen Städte Westdeutschlands und Folgerungen für die regionale Planung und Steuerung*, Arbeitsmaterial Akademie für Raumforschung und Landesplanung Nr. 262, Verlag der ARL, Hannover

Aring, J. (2004) *Stadtregionen: Neue Herausforderungen, neue Aufgaben*, discussion paper, http://bfag-aring.de/pdf-dokumente/Aring_2004_Saar_Vortrag.pdf
Bayerischer Städtetag (2003) *Reform der Gewerbesteuer – Anforderungen und Auswirkungen: Ein Modell des Bayerischen Städtetages*, München
Bayerisches Landesamt für Statistik und Datenverarbeitung (2004) *Gemeindedaten 2003*, München
Bayerisches Landesamt für Statistik und Datenverarbeitung (2005) Datenbank GENESIS online, https://www.statistikdaten.bayern.de/genesis/online/Online, 29 May 2005
Bayerisches Staatsministerium für Landesentwicklung und Umweltfragen (2003) *Arbeitshilfe Kommunales Flächenressourcen-Management*, München
Bayerisches Staatsministerium für Landwirtschaft und Forsten (2004) *Land- und Forstwirtschaft in Bayern. Daten und Fakten 2004*, München
Bayerisches Staatsministerium für Wirtschaft, Verkehr und Technologie (2002) *Der Flughafen und sein Umland – Grundlagenermittlung für einen Dialog, Teil 1: Strukturgutachten*, München
Bayerische Staatsregierung (2003) *Landesentwicklungsprogramm Bayern*, München
Boustedt, O. (1970) Zur Konzeption der Stadtregion, ihrer Abgrenzung und ihrer inneren Gliederung – dargestellt am Beispiel Hamburgs, in Akademie für Raumforschung und Landesplanung (ed.) *Zum Konzept der Stadtregionen – Methoden und Probleme der Abgrenzung von Agglomerationsräumen*, Forschungsberichte des Ausschusses „Raum und Bevölkerung" der Akademie für Raumforschung und Landesplanung, Hannover, 13-42
Brake, K., Dangschat, J.S. and Herfert, G. (2001, eds.) *Suburbanisierung in Deutschland – aktuelle Tendenzen*, Leske und Budrich, Opladen
Bundesamt für Bauwesen und Raumordnung (2003, ed.) *Siedlungsstrukturelle Veränderungen im Umland der Agglomerationsräume*, Forschungen Heft 114, Selbstverlag des Bundesamtes für Bauwesen und Raumordnung, Bonn
Bundesamt für Bauwesen und Raumordnung (2004) *Darstellung des Themas Siedlungs- und Verkehrsflächenentwicklung auf der BBR-Homepage*, http://www.bbr.bund.de, download 24 November 2004
Bundesamt für Bauwesen und Raumordnung (2005) *Mengensteuerung der Siedlungsflächenentwicklung durch Plan und Zertifikat, Informationen zur Raumentwicklung*, 4/5, Selbstverlag des Bundesamtes für Bauwesen und Raumordnung, Bonn
Burdack, J. and Herfert, G. (1998) Neue entwicklungen an der peripherie europäischer Großstädte, *Europa Regional*, 2, 26-44
Commission of the European Communities (1999) *ESDP – European Spatial Development Perspective: Towards Balanced and Sustainable Development of the Territory of the European Union*, Office for Official Publications of the European Communities, Luxembourg
Commission of the European Communities (1999a) *The EU Compendium of Spatial Planning Systems and Policies – Germany*, Office for Official Publications of the European Communities, Luxembourg
Einig, K. (2003) Baulandpolitik und Siedlungsflächenentwicklung durch regionales Flächenmanagement, in Bundesamt für Bauwesen und Raumordnung (ed.) *Bauland- und Immobilienmarktbericht 2003*, Selbstverlag des Bundesamtes für Bauwesen und Raumordnung, Bonn, 111-140
Federal Office for Building and Regional Planning (2001) *Spatial Development and Spatial Planning in Germany*, Bonn
*Frankfurter Allgemeine Sonntagszeitung* (2003) Es gibt kein Ende der Welt mehr, *Frankfurter Allgemeine Sonntagszeitung*, 6 July, 27

Göddecke-Stellmann, J. and Kuhlmann, P. (2000) *Abgrenzung der Stadtregionen – Gesamtdeutsche Abgrenzung der Stadtregionen auf Basis der Pendlerstatistik der sozialversicherungspflichtig Beschäftigten,* Bonn

Heitkamp, T. (2002) Motive und strukturen der Stadt-Umland-Wanderungen im interkommunalen Vergleich, *Forum Wohneigentum,* 1, 9-14

Hesse, M. and Schmitz, S. (1998) Stadtentwicklung im zeichen von auflösung und nachhaltigkeit, *Informationen zur Raumentwicklung,* 7/8, 435-453

Ismaier, F. (2002) Strukturen und motive der Stadt-Umland-Wanderung, in F. Schröter (ed.) *Städte im Spagat zwischen Wohnungsleerstand und Baulandmangel,* RaumPlanung spezial 4, 19-29

Kagermeier, A., Miosga, M. and Schußmann, K. (2001) Die region München – auf dem weg zu regionalen patchworkstrukturen, in K. Brake, J.S. Dangschat. and G. Herfert (eds.) *Suburbanisierung in Deutschland – aktuelle Tendenzen,* Leske und Budrich, Opladen, 163-173

Krau, I. (2005) Stadtregion als kooperatives Netzwerk. Mobilität und Kommunikation am Beispiel Münchens und seiner Region, *Raumforschung und Raumordung,* 1, 47-54

Landeshauptstadt München (2002) *Raus aus der Stadt? Untersuchung der Motive von Fortzügen aus München in das Umland 1998 – 2000,* München

Landeshauptstadt München (2004) *Statistisches Jahrbuch der Stadt München 2004,* München

Landeshauptstadt München, Referat für Arbeit und Wirtschaft (2005) *Standortinformationen,* http://www.muenchen.de/Wirtschaft/Wirtschaftsstandort/Standortberatung/foerderung/standortinfo/79024/index.html

McKinsey, Stern, ZDF and AOL (2005) *München als Investitionsstandort Spitze in Deutschland,* Pressemitteilung 28 April 2005, http://www.perspektive-deutschland.de/files/ presse_2005/pd4-PM_Muenchen.pdf

Mönnich, E. (2005) Ruinöse Einwohnerkonkurrenz. Eine Analyse von Suburbanisierungsproblemen am Beispiel der Region Bremen, *Raumforschung und Raumordung,* 1, 32-46

Planungsverband Äußerer Wirtschaftsraum München (2002) *Datenspiegel 2002,* München

Planungsverband Äußerer Wirtschaftsraum München (2005) *Regionsdaten Region München 2004,* http://www.pv-muenchen.de/leistung/veroeff/daten/REG_2004.pdf

Regionaler Planungsverband München (2001) *Informationen über die Region in Kürze,* http://www.region-muenchen.com/region/region.htm

Regionaler Planungsverband München (2001b) *Regionalplan der Region München,* http://www.region-muenchen.com/regplan/rplan.htm

Rohr-Zänker, R. (1996) Neue Zentrenstrukturen in den USA: Eine Perspektive für dezentrale Konzentration in Deutschland?, *Archiv für Kommunalwissenschaft,* 35(2), 196-225

Statistisches Bundesamt (2005) *Monatsdaten Arbeitslosenquote,* http://www.destatis.de/indicators/d/arb210ad.htm, 30 May 2005

Chapter 4

# Residential Growth and Economic Polarization in the French Alps: The Prospects for Rural-Urban Cohesion

Nathalie Bertrand and Emmanuelle George-Marcelpoil

**Introduction**

Processes involving the concentration of population and economic activities today constitute a major and long-lasting trend in European spatial structures, as demonstrated by the pentagon of economic growth and urbanization outlined by the five cities of Paris, Milan, Munich, Hamburg and London. On a national scale, and from a spatial viewpoint, these polarization processes manifest themselves differently in Europe, but generally raise questions about relations between cities of whatever size and nearby rural areas. Due to possibilities of daily commuting from rural areas into proximate cities (a process that is associated with peri-urbanization), urban expansion occurs and functional relations are established between urban centres and peripheral rural zones within city-regions. In the European Spatial Development Perspective (ESDP, Potsdam, May 1999), these urban-rural inter-changes are considered major issues of policy concern at various geographical scales, with interactions at the city-region scale important for economic and social cohesion, as well as for the sustainability of city-region development (Commission of the European Communities, 1999).

France is concerned by these relationships. Urban expansion and related peri-urbanization processes, which started in the 1960s, constitute major factors in the national landscape. At least 40 per cent of jobs are located in urban zones or peri-urban communes[1] that fall under the influence of an urban zone (Le Jeannic, 1996) and over 75 per cent of the 1999 population was found there (INSEE Première, 2000), a percentage that continues to grow in step with new areas falling under urban spheres of influence. Major features in these processes are the attraction of urban centres (INSEE, 1998), the fragmentation and spreading of urbanized areas (Beaucire and Saint-Gérand, 2001), increased land consumption due to demand for individual homes (Lacaze, 2002) and the saturation of transport infrastructure (Clément et al., 2001). The tensions that result from these processes affect land-use

---

1  The commune is the lowest administrative unit in France (NUTS5 areas).

decisions, social segmentation and the polarization of jobs, which raises doubt as to the sustainability of development in these zones. For this chapter, a critical question is, within this framework of urban dynamics, what is the situation for rural areas?

There is currently debate on the characteristics of contemporary cities and their diversity (Chalas, 2000). In a clear departure from former compact structures for cities, an essential factor of modern cities is urban sprawl and spatial discontinuity. Hence, we find in the literature reference to 'diffuse' cities (Secchi, 2000), that are disseminating urban lifestyles into rural areas, of 'emerging' cities that integrate the countryside and nature (Dubois-Taine and Chalas, 1997), and of 'countryside cities', with this concept highlighting the mutual dynamics of the urbanization of the countryside and the ruralization of cities (Donadieu and Fleury, 2003). As a result, far from opposing each other, rural areas and cities increasingly mix, thus producing new urban forms that take over ever larger parts of undeveloped land. But what are the dynamics of integration between rural and urban areas? Further, as noted in one of the ESDP's political objectives, is there a '... partnership between towns and cities of every size and their surrounding countryside' (Commission of the European Communities, 1999, p.25)?

In France, relations between town and country have long been perceived as being in opposition. Yet the latest laws concerning territorial planning (Chevènement, 1999; LOADDT, 1999) and urban renewal (SRU, 2000) tend to take another approach. They stress the need to reconsider territorial borders by reinforcing inter-communal cooperation.[2] In this context, perhaps we may speak of a new relationship between urban and rural areas, although we also have to ask to what extent the new relationship contributes to the sustainable development of city-regions? Although changes caused by peri-urban dynamics are well-known, there remain questions, in light of these territorial reconfiguration processes, concerning the rural- urban tensions that city-regions continue to provoke.

Consequently, this chapter will address changes in rural-urban relationships and the degree to which new relationships meet the partnership goal contained in the ESDP. It is through the prism of this particular analysis that we intend to review the dynamics currently underway in French peri-urban areas. Two mid-sized city-regions in the Rhône-Alpes NUTS2 region will serve as illustrations. These two will be used as the basis for a more general discussion of development issues in peri-urban areas. The city-regions themselves are Annecy (217 000 inhabitants) and Valence (293 000 inhabitants), which is on the southern edge of the Sillon Alpin (the valley running from Geneva to Valence). The Sillon Alpin is located along the border of Italy and Switzerland. This part of the Alpine Crescent is home to approximately 1.72 million inhabitants. It is one of the most economically attractive areas in Europe, with high-tech businesses, a very dynamic tourism

---

2  LOADDT (1999) targets sustainable development and is based on territorial projects. The Chevènement Law (12 July 1999) targets the reinforcement and simplification of inter-communal cooperation. The SRU Law (30 December 2000) deals with urban planning, city policy, housing and transport. It seeks to set up an approach to urban zones based on solidarity and urban renewal.

sector and a favourable position as a door to Europe in both residential and economic terms (Briquel, 2001). The Sillon Alpin is nonetheless confronted with the problem of maintaining its attractiveness. In coming years, its continuous conurbation will be faced with saturated transport infrastructures, which will be difficult to expand due to a mountainous topography. Both city-regions have diversified economies and major lines of communication (normal and high-speed trains, highways, etc.), which impact on land-use, and encourage population exchanges (recent bilateral agreements have seen many Swiss migrants arrive in Annecy, while Valence has population exchanges with Lyon and Paris).

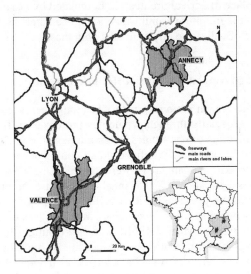

**Figure 4.1 The Location of Annecy and Valence**

Initially in this chapter, we will review the French context of territorial reconfiguration via inter-communal grouping and operations above the communal level, even though relations between communal and inter-communal levels continue in city-regions. Given the absence of inter-communal structures in development plans for city-regions, Annecy and Valence highlight difficulties in spatial functioning. The progressive and functional integration of peripheral areas in these city-regions raises questions for the sustainability of the process and for rural-urban relationships. In the second part of the chapter, analysis of this integration will deal with the economic autonomy of peri-urban areas. Their autonomy is significantly reduced by the concentration of jobs in urban cores. This is exemplified by exploring irregularities in the geography of employment and its concentration, primarily in the central city, but also in the hinterland, due to the emergence of secondary centres. The Annecy and Valence city-regions temper, but do not negate, this analysis, due to the distribution of local services. The analysis of residential dynamics then shows that this concentration is contradicted by the spread of homes in response to demand for individual lodgings. Yet, this growth cannot mask imbalance between housing availability and unsatisfied demand, in difficulties becoming a home owner and in a lack of low-income housing. Residential dynamics emphasize urban fragmentation at the commune level and land consumption. Increased demand for transport, scattered building and land-use decisions in communes all raise sustainability issues for city-region development. Yet rural-urban partnerships are present for environmental services. These are largely based on cost pooling, involving inter-communal cooperation. The latter enables communes or peri-urban inter-communal structures either to cooperate with the central city or to maintain a degree of operational independence. Finally, the question of land management and the preservation of living conditions raises debate about the place of agriculture in areas dominated by urban forces, as well as raising issues about agriculture's multi-functional character, its land consumption and the emergence of collective efforts to deal with these.

## Peri-Urban Change Amidst Territorial Reconfiguration

Diversity within the French national territory has long been seen in terms of rural-urban opposition, especially through typologies produced by INSEE (the Institut National de la Statistique et des Études Économiques or French national institute for statistics). INSEE traditionally defined urban communes using population and land-occupation criteria. Urban areas were taken to be communes or sets of adjacent communes with at least 2 000 inhabitants in a continuous built-up area, where the built-up area represented at least half the total population in the specified communes. Rural communes had fewer inhabitants or a less densely concentrated population. As a result, when INSEE used to distinguish urban and rural communes, this was based primarily on the density of buildings and population. The meant the national territory was characterized as having rural and urban area that appeared as if in opposition to one another. This representation was overly simple. It ceased to be relevant when tendencies for rural population losses becoming city gains changed into rural growth associated with urban sprawl and peri-urbanization. Rural areas now house a new population category, of residents who work in urban areas but live in the 'periphery'.

### Beyond Traditional Rural-Urban Opposition

Observations concerning the increasingly fuzzy border between rural and urban areas led to a reconsideration of the national territory, so it is no longer seen in terms of rural and urban areas, but in terms of spheres of influence. Communes are now ranked according to the intensity of commuting to the closest city, with three area-types resulting (INSEE/INRA, 1998). The first type is made up of urban centres in which there are at least 5 000 jobs. The second is comprised of zones from which commuters are attracted to urban centres, thereby including peri-urban rings from which at least 40 per cent of the economically active population works in the urban centre (this area-type includes communes sending this volume of its workforce to more than one urban centre). The third area-type is predominantly rural, with four sub-types identified, depending upon their functioning as job centres and their relationship with a city.

Above and beyond describing the national territory, the latest territorial planning and urbanization laws attempt to transcend notions of a rural-urban opposition. Hence, the Law on Territorial Planning and Sustainable Development (LOADDT)[3] speaks of '... complementarity and solidarity between rural and urban areas' (Article 21) and mentions territories of differing sizes located between areas traditionally identified as 'rural' and 'urban' (Michelangeli, 2002). The SRU Law[4] on urbanization follows suit when it speaks of '... sparing and balanced use of natural, urban, peri-urban and rural land ...' (Article 1). Similarly, the new law on the development of rural territories (DTR, 24 February 2005) explicitly mentions the 'preservation and valorization' of peri-urban areas. There is then no shortage of legislative acknowledgement of the need to transcend opposition between rural and urban areas.

Finally, these laws, and the Chevènement Law (12 July 1999), target the reinforcement and simplification of inter-communal cooperation, and encourage the emergence of 'relevant' territories, which should be sized to take into account the 'spatial contexts of current problems' through inter-communal policy. Inter-communal arrangements are presented as a means to correct territorial fragmentation (there are more than 36 000 communes in France) and are seen to go beyond rural-urban opposition to provide a territorial organization based on cohesion (Goze, 2000). This is to be characterized by the voluntary participation of communes in communautés de communes[5] and communautés d'agglomération.[6]

## Political and Geographic Reconfiguration in City-Regions

Even if there is no legal recognition that rural-urban opposition is a thing of the past, the reality of the situation is seen in terms of urban influence on surrounding rural communes and inter-communal cooperation. Inter-communal structures are launched locally, on a voluntary basis, with national funding incentives. They now cover a wide range of activities. Depending on region, inter-communal structures are more or less developed. In city-regions, they are notable for undertaking

---

3  Law number 99-533, dated 25 June 1999, *Official Bulletin* dated 29 June 1999, pp. 9515 and following.
4  Law number 2000-1208, dated 13 December 2000, *Official Bulletin* dated 14 December 2000, pp. 19777 and following.
5  An inter-communal local-government entity supplying collective services to a number of participating communes.
6  Designated for an urban zone, by law these are responsible for economic development, spatial planning, the social balance of housing and urban policy. They must also assume responsibility for three of the following five sectors: roads, wastewater, water supply, environmental protection, and large cultural and sport installations. Mandatory responsibilities are less than those required of a communauté urbaine, with this type of structure generally set up for small cities and surrounding areas.

20-year planning and development projects. Although these SCOT[7] plans are theoretically mandatory for all cities with over 50 000 inhabitants (SRU Law, 2000, modified by the HU Habitat and Urbanism Law, 2003), they have encountered difficulties. Opposition often develops between the central city and areas in the periphery of a city-region, where communes and inter-communal structures want a degree of autonomy (Bertrand and Marcelpoil, 2001). When inter-communal structures are set up locally, they reconfigure rural-urban relationships, with, on the one hand, a communauté d'agglomération and, on the other, peri-urban inter-communal structures, which are often limited to a small number of urban communes and rural communes under urban influence.

Our two study areas, like others in France, provide a sign that, despite legislative mandate, policy is not always implemented. Most visibly the signal for this is that neither the Annecy nor the Valence-Romans city-regions have yet succeeded in setting up a SCOT, even though attempts have been made. In the case of the Annecy city-region there is dynamism in inter-communal cooperation, which has resulted in no less than 10 inter-communal structures being authorized to raise taxes. These councils cover almost the entire city-region. The urban core of Annecy, with a communauté d'agglomération created on 1 January 2001, is a sizeable inter-communal cooperative structure in its own right, with its 13 constituent municipalities (i.e. communes) encompassing a large land area. This inter-communal structure encompasses 130 000 inhabitants according to the last census. Inter-communal structures around the Annecy urban agglomeration are characterized by in-fighting among elected officials. Details on partisan rivalries are not important. Suffice it to say that these cooperative structures were created in reaction to those in the Annecy urban agglomeration. They have contributed to blocking actions involving themselvers and municipalities of the urban core.

In the Valence city-region, the dynamics of inter-communal cooperative structures vary. The Valence Inter-Communal Syndicate (or SISAV, which is known as Valence Major), comprises only seven municipalities. This group started in the 1990s to manage public transport, waste disposal and economic matters jointly. In parts of the periphery, there is a longer history of inter-communal co-operation. For the urban centre of Romans, this goes back to a solidarity and co-operation movement developed by the Catholic agricultural youth organization (JAC) for farmers on the Romans Plain. Today, six inter-communal structures and the Valence Major raise their own taxes. They cover a great part of the city-region and often range well beyond it. Yet many municipalities in the centre of the plain do not participate in these inter-communal co-operations.

## Territorial Dynamics, Peri-Urbanization and Functional Integration

At the national level, continental France recorded an annual population growth rate of 0.58 percent between 1999 and 2003, with this level not having been achieved since the growth years of 1945 to 1975. Growth has risen from 0.39 per cent for

---

7  SCOT, schéma de cohérence territoriale (coherent territorial plan).

1990-1999 (Piron, 2005). This growth was largely due to internal expansion (75 per cent), with only one-quarter due to in-migration (INED-EUROSTAT, 2003). The territorial distribution of this growth has two foci. These are small rural communes with low population densities and communes with more than 100 000 inhabitants.

Peri-urbanization processes are active in this context and may be analyzed as gradual processes of urban integration. This integration and the strength of the linkages to towns can be expressed using different analytical criteria, such as geographical proximity, economic ties, institutional arrangements and the supply of services to inhabitants (Briquel and Collicard, 2002). The tightening of linkages associated with peri-urbanization processes is accompanied by social and economic change, as expressed through the arrival of new residents in rural areas, increased demand for urban services, changes in how land is occupied and increased transport demand. Some changes are generic to peri-urbanization (at least at the national level), whereas others are noteworthy in specific regional contexts. In this regard, French housing policy in the 1970s played a considerable role by encouraging the ownership of individual homes (Jaillet, 2004). As a result, homes were built in massive numbers, at ever increasing distances from central cities. With the consumption of housing increasing as incomes rose, while a drop in transport costs helped make possible a rise in individual motorized vehicles (Cavailhès and Selod, 2003), peri-urban communes grew rapidly from 8.9 million inhabitants in 1990 to 12.3 million in 1999. By this latter date they covered 33 per cent of the national territory (INSEE Première, 2000). The spatial extent of city-regions is therefore a major phenomenon today. This phenomenon has been accompanied by the gradual economic and social integration of new rural areas into city-regions. But is it possible to speak of consistency and sustainability in the light of the distribution of jobs, homes and services?

## Job Concentration-Deconcentration and Spatial Cohesion

In France population growth is occurring in both rural areas and in densely settled urban centres, which is a singular occurrence in Europe (Piron, 2005). Yet economic activities are increasingly concentrated. In 1999, 354 urban centres with at least 5 000 jobs each accounted for 72 per cent of paid-employment but only 16 per cent of the national territory (Hilal and Schmitt, 2003). Half of the towns with 10 000 to 20 000 inhabitants were included in these 354, as were virtually all towns with at least 20 000 inhabitants. Above and beyond any general observation concerning the polarization of economic activities, the dynamics of job locations within city-regions need noting. Many studies of urban areas raise questions concerning the concentration of economic activities (Velz, 1993; Pumain, 1994), for they identify opposing forces, both centripetal (with attractions from grouping, such as increased out-sourcing possibilities) and centrifugal (which are often associated with high urban costs for real-estate, transportation, wages, etc.), with the playing out of these forces in specific contexts explaining final spatial configurations (Krugman, 1996; Cavailhès and Schmitt, 2002). Resting 'on top' of these forces, there is a trend in metropolitan regions toward a redistribution of jobs between the central city and its suburbs. Thus, in North American metropolitan regions, the decreasing density of jobs and their spatial reconcentration have resulted in the emergence of new centres and a recomposition of metropolitan territories. North American studies note the increasing autonomy of 'suburban' areas, with new job centres created or coming to the fore on the periphery of metropolitan regions (Bourne, 1996). This redistribution of jobs has modified intra-metropolitan relations. Many municipalities in the older suburbs now have more jobs than the central city, such that, whereas two-thirds of jobs were in central cities in the 1960s, today only 18 per cent are within five kilometres of the centre (Lacaze, 2002). What is more, starting in 1990, more people commuted to work from one suburb to another than from suburb to central city. For some commentators there have been distinct steps in this spatial restructuring, starting with bedroom suburbs (1950-1965), then a greater residential function for peripheral areas due to shopping centres (1965-1975), followed by stronger growth in jobs (1975-1985) and a certain maturity since 1985 with the development of new job centres (Harthorn and Müller, 1989, Giuliano and Small., 1991). A Californian model seems to have announced a general dispersion of jobs (Gordon and Richardson, 1996), with Canadian metropolitan regions also tending toward a polycentric model of job polarization.

In France, a number of empirical studies have shown that job deconcentration is resulting in employment growth along the peripheral edges of urban zones (Baumont and Bourdon, 2002). During 1975-1999, the spread of jobs in the Paris region took place faster than that of population, with the average distance from the centre of Paris increasing by 16.7 per cent for jobs and by 13.5 per cent for the population (Fouchier, 2002). More generally, deconcentration has affected consumer services in notable ways and, due to scale economies, has resulted in the creation of secondary service centres in peri-urban areas (Goffette-Nagot and Schmitt, 1999). Industrial and high-end commercial jobs are also affected, but to a lesser degree. The dominant employment sectors in peri-urban (and rural) areas today are in consumer services, especially those directed at the local population, with such jobs representing 40 per cent of all non-agricultural jobs. These services include stores, commercial consumer services and administrative services (education, health and social affairs). As noted by Vanier (1999), commercial consumer services are dominated by large supermarkets, of which 85 per cent are in peripheral areas. Yet high added-value services remain concentrated in urban zones. With consumer services dominant in peri-urban areas, economic processes impacting on these zones have been termed 'residential' (Hilal and Schmitt, 2003). There has been a deconcentration of jobs, but this is 'segmented', albeit with a multi-centred patterning in areas under urban influence.

To a certain degree, mid-sized city-regions such as Annecy and Valence comply with the above observations. The Annecy city-region is characterized by a double flow of jobs, with concentration in the central city and an outward flow to nearby communes. A few job centres[8] of different sizes structure the peri-urban

---

8   A job centre is any commune with at least 1 500 jobs (Briquel and Collicard, 2002).

zone around Annecy. These secondary urban centres are less new centres than old centres that have taken on greater importance. To take one example, the Rumilly area, which consists primarily of the town of Rumilly (11 617 inhabitants in 1999), is the product of its industrial history. Since the 1960s, Rumilly has been known for its tanneries,[9] with these factories employing a sizable Turkish immigrant population. These tanneries attracted a highly qualified labour force, which then drew in companies from outside the area. These companies have since become major actors in the local economy (e.g. Nestlé for agro-food processing). These inflows reinforced the area's industrial image, which is based essentially on production units. That being said, most of the company headquarters are located in the Annecy urban area, which is a sign of a certain economic dependence.

Other communes in the south of the city-region are under the economic influence of centres outside the zone or on the northern edge of the study zone, next to the Swiss border. Whereas in larger cities it is common for commuting flows to radiate from main centres, a characteristic of the Annecy City-Region is commuting to Switzerland. These jobs account for between 10 per cent and 24 per cent of the resident workforce in communes closest to the border. These linkages, affect all communes in the city-region but drop off quickly with increasing distance and travel times from the border. Also of note is the fact that the deconcentration of jobs from the central city has not been associated with economic specialization, with a few exceptions. Annecy is affected by this trend. However, a number of shopping centres are emerging on the outer edges of the city.

Unlike Annecy, the Valence city-region has multiple urban centres; namely, Valence, Romans and Tain-l'Hermitage. Valence has the strongest pull in terms of jobs, whereas the attraction of the two other centres does not go beyond peripheral communes. The economy of this city-region was originally structured around these three major centres, which have diverse traditions and know-how (electro-mechanical devices, leather and shoes, jewellery, metalworking, etc.). However, difficulties encountered since the 1960s (the first a slump in the shoe market for Romans in 1963, with a crisis in the building sector in Valence at the beginning of the 1970s) contributed to a recent dilution of territorial specificities and consequently competition between territories. Romans is still a major centre in the leather industry, with the presence of large French shoe brands. Today, these centres are characterized by job deconcentration to areas along major, existing highways. Generally, companies are leaving the urban zones (a number of causes are mentioned, notably the level of company taxes) and setting up in peripheral areas, but still with a north/south orientation (along the A7 highway).

## Local Government and the Spatial Structuration of Jobs

Debate on the dynamics of job reconcentration or dispersion in city-regions raises a number of issues concerning public action and local regulation. For example, in Canadian metropolitan regions, the political role maintained by the central city is

---

9   Today, only one remains.

based largely, above and beyond diverse economic dynamics, on public management of networks (Manzagol and Coffey, 2001). More specifically, in France, certain types of infrastructure or major services, such as medical laboratories and particularly hospitals, tend to concentrate in heavily populated communes. Public authorities may play a role in maintaining or developing these non-commercial services (Hilal and Schmitt, 2003), depending on local synergies between public and private initiatives. Public intervention is widely mentioned in maintaining territorial equity, but little mention is made of the influence of inter-communal structures or even of their emergence when the spatial dynamics of jobs is explored. This prompts the question, does inter-communal cooperation, which after all is mandatory in France in economic development promotion (Massini, 2001), play a role in the polarization or dispersion of jobs?

Negotiations for 20-year spatial planning and economic development projects contained in a SCOT generally seek to balance housing with jobs, given these are recognized as major issues for sustainable development in city-regions (Martin et al., 2004). The city-regions of Grenoble (626 000 inhabitants in 1999) and Chambéry (205 000) are examples of such efforts within the Sillon Alpin (Bertrand et al., 2005). These city-regions are just a few dozen kilometres from Annecy and Valence. They have initiated wide-ranging inter-communal cooperations (involving 202 and 103 communes, respectively) at the scale of the city-region, and between the urban centre and surrounding communes. For France, these two city-regions constitute outstanding examples, as inter-communal cooperation is often found to be difficult, even where municipalities are much smaller. This is the case in Annecy and Valence, where there are peri-urban communautés de communes with 10-15 communes or fewer, and a communauté d'agglomération (comprised of Annecy, with 10 communes). In these city-regions, there is no planning and development project for the entire city-region within the framework of a SCOT. In this context, a number of observations must be made on the polarization of jobs.

In the Annecy city-region, over 65 per cent of manufacturing jobs, and 70 per cent in certain services, are in the urban centre. The structure in communautés de communes and in the communauté d'agglomération reflect these differences rather than reducing them. The groups of communes that exist make the spatial imbalance in jobs more rigid, as well as consolidating opposition between rural and urban communes. For 2000, the Annecy communauté d'agglomération provided almost 65 per cent of manufacturing jobs, 86 per cent of services for companies, 73 per cent of (wholesale and intermediate) trade and 83 per cent of health and social services (Unedic).[10] Nine of the 10 communautés de communes generally possess less than 5 per cent of jobs in each of these sectors. The only exception is the Rumilly communautés de communes, which is centred on the second urban centre in the city-region, with 20 per cent of manufacturing jobs. Moreover, the major housing and jobs imbalance that exists in the hinterland is worsening. This is

---

10 Data provided by Unedic, which is the Union Nationale pour l'Emploi dans l'Industrie et le Commerce (the national union for industrial and commercial employment). These computations were made by D. Borg at Cemagref.

raising pressing questions concerning the risk of communes becoming little more than bedroom communities, while local road networks are becoming saturated.

Set amidst this condition, the planning and management of business parks at the inter-communal level has met with difficulties. The Annecy city-region is largely characterized by the efforts of individual municipalities to create business parks. Very few business zones result from a coordinated policy, although the Allonzier-la-Caille business park, in the north of the city-region, is an example of one inter-communal park. More recently, further projects have come into existence. For example, in the west of the city-region, around the town of Alby-sur-Chéran, an inter-communal zone has been set up by combining existing municipal zones. This new business park has a strong economic base in transport and logistics, and is becoming the logistics platform for the Annecy urban agglomeration. A large business park is also being created in the Annecy communauté d'agglomération, in an effort to specialize in commercial and service activities.

Unorganized efforts to create business parks result in 760 hectares being available for sale and the installation of companies in the Valence city-region in 2000. This corresponds to 30-60 years of demand.

But while job imbalances maintain opposition between central cities and peripheral areas, in-depth analysis of service densities leads to a more nuanced view, with the exception of services for companies and, to a lesser degree, health. For services for companies, the employment density in the communauté d'agglomération is higher (79.2 jobs per 1 000 inhabitants in 2000) than in peri-urban communautés de communes (less than 15 jobs per 1 000 inhabitants). Health and social services follow the same trend, with 24 jobs per 1 000 inhabitants in the urban centre, compared to less than 9 jobs per 1 000 inhabitants in most peri-urban communautés de communes. On the other hand, (wholesale and intermediate) trade in the peri-urban communautés de communes is 1.5 times higher than in the communauté d'agglomération, which has only 27.4 jobs per 1 000 inhabitants. Finally, densities for retail and local stores, while fairly low, are higher in certain peripheral communautés de communes. The figures do not single out the urban centre as a focal point for jobs, for proximity is maintained between the population and certain services (Hilal and Schmitt, 2003). It follows that when discussing jobs and peri-urban economic development, it is necessary to distinguish two scales (Rousier, 2001). On the one hand, there is the productive city-region economy, with large business parks in outlying areas. Its importance extends to all institutions in the city-region participating in economic development. On the other hand, there is a residential economy with local service demands that must be met.

## Peri-Urban Dynamics and the Trend Toward a Residential Economy

'Growth in rural communes therefore represents half of the total population growth in continental France between 1999 and 2004' (Piron, 2005, p.4). This growth pattern constitutes a departure from the traditional image of French rural areas losing population to urban centres. Growth rates recorded between 1999 and 2003 differentiate between various areas and highlight rural dynamism. Rural communes have an annual growth rate of 0.68 per cent whereas 'predominantly urban areas' have an overall rate of 0.56. Urban centres grew less quickly with 0.32 per cent. The two city-regions studied in this chapter fit into this general trend (Table 4.1).

**Table 4.1   Population Characteristics of the Urban Region of Annecy**

|  | Population | | | |
|---|---|---|---|---|
|  | Inhabitants 1999 | Density 1999 | % annual growth rate 1990-1999 | 1968-1990 |
| Urban Region Annecy | 216 784 | 218 | 1.30 | 1.98 |
| Urban Zone Annecy | 140 868 | 958 | 0.85 | 1.81 |
| Rural zones | 75 916 | 89 | 2.19 | 2.35 |

Source:   French population censuses for 1968, 1990 and 1999.
Note:   Density refers to inhabitants per square kilometre.

The Annecy city-region has been characterized for the past two decades by the spread of the Annecy urban agglomeration, with major development along the edges of the lake, which is in the process become an urban lake, and fragmented urban development in peripheral areas. An explosion in urban development is nonetheless a relatively recent phenomenon, having started in the period 1975-1980, although economic growth in Annecy is much older. It is possible to distinguish three steps in the city's expansion. The first took place during the 1960s and 1970s, when housing, essentially individual, was built primarily around the lake. During the second phase, which lasted until approximately 1990, growth moved to adjacent hills and slopes, taking advantage of the mountain geography of the region. Then came a form of urban overflow, which is generally considered to be less organized, occurring along roads, most notably the national roads that run north-eastward and south-eastward of Annecy.

**Table 4.2   Population Characteristics of the Urban Region of Valence**

|  | Inhabitants 1999 | Density 1999 | % annual growth rate 1990-1999 | 1968-1990 |
|---|---|---|---|---|
| Urban Region Valence | 293 431 | 202 | 0.58 | 1.00 |
| Urban Zone Valence | 149 431 | 651 | 0.36 | 1.02 |
| Urban Zone Romans | 46 431 | 761 | 0.20 | 0.40 |
| Rural zones | 97 569 | 84 | 1.11 | 1.32 |

Source:   French population censuses for 1968, 1990 and 1999.
Note:   Density refers to inhabitants per square kilometre.

Valence has also experienced growth, with 1950-1975 seeing California-style rapid growth, which doubled the population from 50 000 to 105 000. The various stages of urban expansion occurring within the city-region proceeded virtually in step with different phases of economic development and 1970s collective housing programmes. Reflecting a larger trend, the past few years have been characterized by a change in demand toward individual homes in peri-urban sectors. Prior to 1990, urban expansion took place primarily to the west of Valence. Difficult access to the City, as there is only one bridge over the Rhône, and a limited amount of space afforded by the Vivarais Hills to the west, have resulted in a shift to east of the City. This new expansion should continue until it reaches the Vercors Mountains, with expansion in this direction reinforced by a dense network of high-quality roads. Although the main post-war trends have seen strong north-south growth along the Rhône, a decline in central city populations and the development of peripheral rings, across France recent years have revealed demographic renewal in central cities in urban agglomerations (e.g. Ogden and Hall, 2000), alongside the attraction of rural areas. Valence, for example, recorded a sharp drop in population losses (due in part to increased housing units), while peri-urbanization continued in parallel. While overall population growth brings together rural and urban zones, the question remains as to whether this growth and the resulting demand for housing contributes to sustainable residential development.

**Population Growth, Social Segregation and Housing Access**

Some general observations can be made about social dynamics in peri-urban areas. The peri-urban population is primarily middle-class (young couples with children), for whom the move into a peri-urban zone represents a step up the social ladder, although these areas are far from socially homogeneous (INSEE/INRA, 1998). The middle levels of society have highly diversified workforce engagements and professional occupations. The outward appearance of peri-urban areas is rather that of a mosaic of social categories, which group according to income levels.

The spatial hierarchy of these social categories is relatively well-known in France, and is evinced in indicators of social-professional composition and business activity (Tabard, 1993). However, the causes of social differentiation and its relationship to real-estate markets has been interpreted in different ways (Granelle, 2002). According to sociological hypotheses, real-estate markets reflect segregation processes that derive from members of social groups seeking to live with those of similar social standing. In contrast, hyopotheses from an economic viewpoint suggest that urban segregation results from real-estate markets being driven by economic and cultural factors. In the latter case, these factors are above all economic, whereby households make virtually the same decisions in terms of the balance between real-estate costs and transport costs, depending on social class. Added to these factors, analysts recognize that differentiation is affected by the importance placed on cultural services or landscape amenities (Cavailhès and Selod, 2003). The social mosaic that results can be conceptualized as a product of three overlapping patterns. The first pattern is a ring of concentric circles, whereby the middle classes are distributed from the city centre outward to the peripheral areas according to their household income, with the highest incomes closest to the centre. The second is a linear pattern, as along major transportation lines. The third is a spot pattern, where a particular landscape or the environmental quality of a site enhances its value (Jaillet, 2004). This socio-spatial sorting is part of the broader phenomenon of urban social segregation. In this regard, it is worth noting Donzelot's (2004) vision of a three-facetted urban situation, with the first being the expulsion of poorer populations to large housing projects on the urban edge, the second being city centre gentrification[11] and the third being peri-urbanization.

More generally, social segregation raises the question of the balance between housing supply and demand. Demand for housing is essentially urban in origin and primarily favours purchases of individual homes (Cavailhès and Selod, 2003), most notably by young couples and mid-management executives. This trend is in line with economic analyses of real-estate markets, which reveal a drop in prices as distance increases from the central city. Often strong demand for collective rental units and for low-income housing is far from met (this was the main purpose of the SRU Law, 2000).

In this context, limited housing supply may be seen to result from land shortage, the reasons of which may be analyzed along two lines. The first arises from communal management practices, for the commune has the authority to establish land-use zoning. Although some communes have discreetly called on communauté de communes to manage low-income housing, residential planning is still in the hands of commune authorities. Policies at this level, which are uneven in the way they favour opening land for construction, result in scattered building outcomes, owing to the absence of an overall land-use planning project.

The second factor is the pace of population change. In the Annecy city-region, for example, population change is particularly dynamic, with generally strong growth in the Haute-Savoie département. Thus, for 1975-1990, the urban agglomeration recorded a growth of 2 000 inhabitants a year, which helps explain difficulties in terms of meeting housing demand. Annecy is characterized by insufficient inexpensive housing, a factor that is not without consequences for the local economy. As a result, certain socio-professional categories, notably seasonal workers (a sizeable segment of the workforce as tourism is important) cannot find housing at a reasonable cost. Similarly, companies encounter increasing difficulties finding housing for their workers. This is despite the fact that, during the 1960s, two major housing projects were completed in priority urbanization zones (ZUP or Zone d'Urbanisation Prioritaire), so making subsidized housing available. Today, demand for individual housing units represents 80 per cent of the total. However, the demand for collective housing is strong and is not being answered by political initiatives within the urban agglomeration. Moreover, housing shortage and social

---

11 The term gentrification is rarely used in France, yet this socio-spatial process is at work in French cities. For Bidou-Zachariasen (2004) a factor in the shortage of studies lies in traditions in French sociology, which has promoted a Marxist view of social relations, and finds the middle classes a 'uncomfortable' subject. She speaks of a 'French taboo'.

segregation are expected to worsen as a result of recent bilateral agreements between France and Switzerland that allow Swiss citizens to live in France. This has already increased pressure on real-estate prices.

## Housing Supply and Demand Imbalances and Economic Development

Imbalance between housing supply and demand has created a difficult situation for home buyers and those wishing to rent apartments. The situation is worse for low-income housing, for which availability is far from sufficient. Excess demand has resulted in a major rise in the cost of land, particularly over the past few years (see Table 4.3). In a number of cases, this weighs heavily on the local economy.

In the Annecy city-region, the imbalance between supply and demand exists in parallel with a major increase in real-estate prices due to a shortage of land. Communal policies on opening of land for construction differ even within communauté de communes. Policies are motivated by different factors, including the desire to offer housing for young people in a commune, a wish to maintain a range of social classes, concerns over financial implications (such as the cost of services like water treatment) and plans to integrate newcomers progressively. Policies have also been formulated with an eye on relations with the central city, with some communes not wanting to become bedroom communities for Annecy. Many communes wish to retain a 'rural' identity, although what this means has changed over time, with 'rurality' now seen as a desirable residential category (it may also be viewed as a desire of new residents to exclude certain social classes, thus contributing to the phenomenon of social segregation in the hinterland).

**Table 4.3   Change in Land Prices in Annecy, 1997-2001**

Price of building land per square metre (€)

|  | 1997 | 1999 | 2001 |
|---|---|---|---|
| City-region Annecy | 30.0 | 32.9 | 45.2 |
| Urban zone Annecy | 48.0 | 52.2 | 83.6 |
| Rural zones | 22.7 | 27.1 | 35.9 |

Source:   Rural Land Management Office (SAFER) and notaries data.

Housing changes over the past 15 to 20 years are directly related to a shift toward a higher rate of land consumption, which results from urban pressure on peri-urban communes, the majority of which can still be termed 'rural' (with agriculture occupying large areas). Housing in these communes generally consists of individual dwellings that are the permanent homes of their occupants. Two major trends are associated with these changes. The first involves the rapid release of land for new populations. The second is a desire to control the arrival of new populations. The latter is justified by a number of motivations, such as concern about integrating new populations without conflict, about avoiding indiscriminate building, over maintaining a balance between new arrivals and available services, and about the costs of growth for a commune.

What should also be noted here is that, while communes have some leeway with respect to encroaching urbanization, private land owners generally attempt to make housing plots available, for real-estate speculation is part of land owners' financial strategies. Added to which, local inhabitants can (and do) attempt to limit housing densities, as well as collective housing projects, whether for low-income units or not. Generally speaking though, it is change to planning policies that have slowed land consumption (Morand-Deviller, 2003; Lecat, 2004), and it is still essentially the communes, via their PLU (local urbanization plan), that promote their own land-use policy (Martin, 2001). The splitting of decision-making capacities when there is no SCOT for a city-region raises two questions. The first concerns land consumption, the second the spatial cohesion of such regions and resulting social and economic impacts.

## Inter-Communal Solutions for Urban and Environmental Services

The arrival of new (primarily urban) populations has raised the question of service provision in communes which recently had much smaller populations (Bertrand and Vollet, 2002). Such services have changed over the past 10 years, becoming more diversified as demands have risen from a growing, younger population. More or better stores, commercial services and administrative services (education, health, social affairs) are considered 'normal' by newcomers (Rousier, 2001), even though the peri-urban context, alongside the limited financial capacities of communes and inter-communal structures, make the technical and financial management of services more difficult. Shortfalls in private sector services have resulted in institutional regulations governing certain services being introduced by local government entities with a view of ensuring a balance between the public and private sectors (Hilal and Schmitt, 2003).

Depending on sector, services are provided at the commune or inter-communal level. Many services are provided by private entities or associations that often operate over territories far larger than a single commune (e.g. ADMR, Aide à domicile en milieu rural, which offers home assistance – cleaning, social assistance, meals, etc. – in rural areas, which generally operates at the inter-communal level). The demand for school and school-related infrastructure from newly arrived families with young children is the responsibility of the public sector (day-care centres, school restaurants and schools themselves). This element of service demand has increased considerably over the past few years. However, if services that were traditionally handled at the inter-communal level are excluded, it cannot be said that the management of services has fallen significantly on the inter-communal level. The limited finances of communes has meant that the supply of services has not been sufficient to meet increasing demand. But this has resulted in a shortfall in supply, even though the delegation of commune responsibilities to the inter-communal level might have improved the situation. The one field in which service provision has often allowed inter-communal collaboration between the urban core and peripheral communes is for 'environmental' services.[12]

## Environmental Services at the Inter-Communal Level

Growing legal demands concerning waste and water compliance with European standards have led peri-urban territories to join large 'technical structures' that are capable of producing scale economies that are not available for communes or small inter-communal units. Environmental services for waste management, water treatment and the supply of drinking water, represent a major outlay and require the sharing of costs and the transfer of responsibilities to the inter-communal level. This obligation, reinforced by the environmental standards imposed by the Water Law (1992), leads communes to join forces, although occasionally with difficulty in the light of the autonomy some claim. These groupings do not seem to reinforce the creation of local inter-communal structures, but they are indicative of the financial constraints that environmental management presents for local structures (Vanier, 2000). They also raise the question of whether more comprehensive management is required to address inter-sectoral environmental problems in a more integrated manner.

Water treatment in the Annecy city-region provides a good example of participation in a composite[13] inter-communal council for Lake Annecy (SILA). This was created in 2001 to replace the existing inter-communal council for Lake Annecy. The creation of SILA is thus the product of a older inter-communal culture, plus a political desire to 'stand together' for environmental management. The effort to clean Lake Annecy, which was gravely threatened at the time, presented the initial challenge. Today, SILA is responsible for the collection and treatment of waste water and for garbage disposal. SILA collects and processes household garbage and similar waste for all participating communes and inter-communal structures (including the Annecy communauté d'agglomération). It covers 113 communes and more than 250 000 inhabitants. For over 40 years, an inter-communal structure for Lake Annecy has processed garbage, with a yearly volume increase of 2 per cent. The projected figure for the year 2010 is 155 thousand metric tons. To manage this volume, SILA utilizes an array of techniques, including material recycling via collection sites and in-home sorting, composting, incineration with energy recovery and sophisticated filtering of smoke. The goal is to provide the public with the best possible service while establishing systems and techniques that maintain the quality of the environment.

Examples of inter-communal structures in the Annecy city-region reveal the impact of shared costs for environmental services. Some reveal a wish to maintain political autonomy (Bertrand and Marcelpoil, 2001). Others involve cooperation within the urban agglomeration. Technical environmental services are a common

---

12 These services are called urban services in some publications in as much as they are related to higher housing densities and the urbanization of land.
13 A 'syndicat mixte' is a council whose members are from different types of local government entity.

factor behind inter-communal cooperation, especially for environmental purposes. This is illustrated by the Fier et Usses communauté de communes to the north-west of the Annecy urban agglomeration, which covers a residential expansion zone. For Fier et Usses, water treatment was the major factor in setting up the inter-communal structure, first in the form of a district, then as a communauté de communes. Population increases and new standards established by the 1992 Water Law led the communauté de communes creating a collective treatment system, in some cases for communes with no previous system at all. The opportunity to share construction and maintenance costs for this infrastructure motivated this decision, which affects the independence of each commune to a certain degree, but was absolutely vital given the volume of newcomer arrivals. Following initial opposition, the decision to join SILA was made in the light of a study of the communauté de communes' water-treatment plan, which made clear that the treatment situation was inadequate, with dilapidation and insufficient capacity in certain existing plants. Confronted with population growth and an inadequate water-treatment network, a number of factors worked in favour of participation, namely the cost of investment in treatment plants, opposition from the préfecture to a new plant close to drinking water sources (this zone was later set up as a biotope reserve), and internal opposition to the communauté de communes from certain communes. In the end, negotiations between the president of SILA and the Fier et Usses communauté de communes convinced communes to join the inter-communal structure. Currently, a number of systems co-exist in the Fier et Usses communauté de communes, including connection to the SILA network along catchment-basin lines, autonomous treatment plants managed by SILA and individual treatment.

Above and beyond considerations of finance, mandatory standards (e.g. for old treatment plants) and insufficient capacity, water treatment impacts on communal spatial planning. It is often a decisive element in deciding where to set up zones for new buildings. This technical issue ties in with political aspects of grouping communes into macro-structures (e.g. SILA). This is because, as there is often conflict between communes, it is only through technical considerations, such as location in the same catchment basin, that these problems are solved. Yet smaller communes often point out that limitations are imposed on their residential expansion by Water Law stipulations about water treatment. To place this point in context, communes have been informed that further urbanization will be possible only if a collective water treatment system is established. These requirements have major financial consequences, with feasibility heavily influenced by natural features (for example, when the ground is a very hard, impermeable molasse). Similarly, certain lots are declared as buildable, then cease to be so, depending on water treatment requirements. What is more, some hamlets have collective water treatment, while others do not.

**Peripheral Rural Identity and Rural-Urban Opposition**

Whereas virtually all urban zones of 30 000 or more inhabitants have as many jobs as resident workers, in peri-urban areas the proportion of jobs to resident workers gets as low as 30 per cent (Hilal and Schmitt, 2003). Above and beyond the autonomy of local areas appearing to be strongly determined by urban hierarchy, efforts to maintain a rural identity tend to play an important role in residential expansion for the urban area. Resistance to the idea of becoming a bedroom community and demand for attractive living conditions are important causes of friction between the central city and desires to conserve natural resources and cultural heritage. These exchanges between the centre and peripheral areas in city-regions raise a situation for natural resources that is rather paradoxical. On the one hand, a positive perception in peri-urban zones is a factor in their residential attractiveness. An additional, more collective role, is that of a 'green lung' or a place for urban dweller relaxation, which highlights the need for fiscal cohesion between urban centre and peri-urban territory. On the other hand, a perception exists in peri-urban territories that natural resources may be a means of remaining autonomous from the central city. This is because resources enable communes and inter-communal structures to proclaim their identity and thus attempt to consolidate their relative autonomy within a city-region.

There is opposition between the outlook that views the city-region as a whole, which targets cohesion, and a vision that centres on communes and/or inter-communal structures, where a key goal is greater autonomy. In understanding these different outlooks, the qualities of rural areas is central.

*Maintaining Peripheral Area Autonomy and Rural Amenity Conservation*

The qualities of rural areas extend well beyond agriculture, incorporating rural amenities as natural or human products that a well-defined territory offering specific natural and/or cultural features (OECD, 1996).[14] Amenities are often presented as a comparative advantage for rural territories, which can be decisive in decisions on where to live and in decisions by companies on where to locate (Beuret and Kovashazy, 2002). Amenities also contribute to the quality of local life, which is one criteria for residential selection. Of course, the phrase 'quality of life' is an imprecise notion, which is subject to varied interpretation. It raises questions concerning personal representations and behaviour, alongside the notion of demand diversity. For residents, the local 'environment' would seem to be based on three main factors (Bertrand and Marcelpoil, 2001): first, social vibrancy, as expressed in the development of associations; second, economic vitality, which assists with the improvement of local services as well as reinforcing territorial images; and, finally, landscape, which is tied to the natural environment, so raising questions about the role and position of agriculture, as the guarantor of land upkeep, and its role in rural images and what rurality symbolizes.

But perceptions of living conditions are diverse. Above all they relate to the manner in which elected officials implement and develop these perceptions in relation to development projects and land-use change. Analysis of these processes reveals that communes which select urbanization and economic growth as a development strategy subsequently find themselves faced with major constraints.

---

14 Quoted in Beuret and Kovacshazy (2002, p.15).

For one, insufficiently planned urbanization (i.e. non-controlled) has resulted in geographically indiscriminate construction, which generates increased concern about the quality of 'living conditions'. Scattered construction expansion also raises questions about the reversibility of development decisions. Past decisions limit future options. This is because the concept of 'living conditions' is based largely on the set of amenities available within a territory. These amenities go beyond economic considerations, dealing notably with the cost of and access to housing,[15] as well as drawing attention to the need to understand '... the attractiveness of the area for certain individuals or social groups, as well as why these people like the area and want to be there, above and beyond the simple satisfaction of material needs' (Cairol and Terrasson, 2002, p. 6).

*Agriculture and Landscape Amenities*

Economic and social change involved in peri-urbanization impact on territories and produce profound modifications in land-use, most notably for agriculture. The role of agriculture, in both quantitative and qualitative terms, lies at the heart of change, in large part due to the pressure exerted on the environment. For this economic sector, the task of providing food is becoming secondary, in favour of more or less clearly expressed requests to maintain the land (Beuret, 1997). In the process agriculture has become multi-functional, as was confirmed by Agricultural Law 99-574 (Loi d'Orientation Agricole, 1999). Yet change in agricultural activities creates friction with other land-uses, notably those involved in urbanization, which are characteristic of peri-urbanization. Peri-urban territories are confronted with major land consumption demands that can severely affect living conditions and compromise capacities to attract new inhabitants in the long-term. The existence of tensions between living conditions and continued development raises more general questions over the conditions required for collective action that is capable of defending rural lifestyles and identities; in short on the possibility of autonomy for peri-urban territories.

The national trend is toward fewer farms in the face of urban pressure. Thus, the number of farms in French peri-urban zones (231 000 in 2000) dropped by 35 per cent from 1988 to 2000, which was '... only slightly faster than in rural zones' (SCEES, 2002, p.1). That being said, the quantity of land rendered 'artificial' increased by 14 per cent between 1992 and 2001. However, as noted by SCEES (2002, p.1), '... this development impinges only marginally on peri-urban agricultural potential. From 1988 to 2000, the quantity of agricultural land used dropped by only 3 per cent in peri-urban areas, compared to 2 per cent in rural

---

15 As early as 1978, the Annecy Lake *White Book* stated that the future of agriculture depended in large part on real-estate questions, which were seen as the make or break factor. The goal at that time was to enable agriculture to contribute to balanced development '... by maintaining stable residents in rural communes, by maintaining landscapes both pleasant and welcoming for visitors, without weighing too heavily on the operating expenses of local government entities' (Livre Blanc du Lac d'Annecy, 1978, p.11).

areas and 12 per cent in urban zones. The land abandoned in peri-urban areas is most often taken over by other farms intent on growing in size'. Certainly, in the Annecy city-region, the general trend toward larger farms confirms the need for ever more land to maintain the profitability of farms, with the exception of the Reblochon cheese AOP zone. This increase in farm size is helped when other farmers cannot find a successor. In a general context of insufficient farmland, farmers go far to find new land to meet the economic need to increase the size of their farms. Consequently, there are two types of pressure on real-estate. The first is urban pressure. The second is primarily agricultural, with pressure arising primarily from competition in the Reblochon zone of the Annecy city-region.

In the final analysis, in urban plans farming proposals often reflect a contradiction between agriculture as an economic activity and agriculture as a means of using land (Tolron, 2002). In parallel, the relation with agricultural land is largely based on complex links with landscape, which is an amenity perceived to be the product of a natural environment that has been more or less built-up and transformed. Friction in urban-agricultural relations arises not only over the quantitative use of land, but also in farm practices and their impact on the environment. Identifying the nature of change, the landscape atlas of Haute-Savoie distinguishes eight major landscape types (Ministère de l'Equipement, 1997).[16] These separate, on the one hand, landscapes undergoing change due to recent, diffuse urbanization, which are portrayed as tending toward residential usage and, on the other hand, agricultural zones that, despite problems, remain economically dynamic with their modernized farms.

For approximately 15 years, analysis and evaluation of agricultural practices have changed due to (ecological, socio-economic, etc.) demands concerning the environment (Rapey *et al.*, 2002). These changes reflect new expectations amongst the urban population. Today, a certain number of features that were originally produced by agriculture, such as hedges, paths, traditional architecture, etc., are no longer produced. Agriculture is seen as indispensable for maintaining open landscapes that constitue an important social amenity for development in certain regions. By expanding the function of agriculture beyond production, such amenity contributes to the multi-functionality of the sector. For a number of years, the Common Agricultural Policy has recognized and encouraged this multi-functionality (Rambonilaza, 2002). Subsidies, formerly based on the market price of products, have been replaced by direct payments to farmers. This recognizes that farmers now provide non-commercial services to society and must be compensated for those services. In France, the Agricultural Law of 9 July 1999, along with the recent Rural Law (DTR, 24 February 2005), acknowledge the multi-functionality of agriculture and seek to organize it via a number of tools.

Generally speaking, amenities generate two critical frictions. First of all, there is an inherent contradiction in public goods being managed on private land or managed by actors with a specific production goal. Secondly, urban-rural relations,

---

16 Natural landscapes, rural landscapes, tourism landscapes, mixed landscapes, residential landscapes, urban landscapes, rural landscapes undergoing change, and landscapes in general undergoing change.

bearing in mind the difficulty of using that expression, produce conflicting interests and legitimacies, particularly concerning land-use and relations between urban and rural inhabitants.

## Local Authorities Intent on Maintaining Quality Living Conditions

The quality of life is also related to how elected officials implement and prioritize development projects and land-use change. Some communautés de communes in the Annecy city-region want to maintain their 'rural' identity, notably through the presence of agriculture. Here the decision to create communautés de communes was largely based on a desire to maintain the current rural character, so as not to become a suburb of the city. Indeed, '... action to support agriculture and contribute to maintaining agricultural structures that are important for the commune or the communauté' is one responsibility assigned to communautés de communes (Communauté de Communes du Canton de Rumilly *Intercommunal Bulletin*, 5, 2002, p.1).[17] To maintain a rural character, it is necessary to preserve land for rural activities, with agriculture (and woodland) often judged synonymously with 'rural'. However, with a landscape essentially characterized by grazing on grassland (dairying), grain cropping is increasingly prevalent, even beyond areas affected by urban pressures, thus modifying the landscape people want to defend.

The existence of farms and land-uses linked to them are perceived as elements in the rural identity of communautés de communes. These two facts are integrated in speeches by local politicians on the rural identity of peri-urban areas, which is seen as a main element in local attractiveness. Yet management of living conditions in peri-urban territories raises questions over the presence of agriculture in these zones. The fall in the farm population, even if the number of jobs is still high in most communes, contributes to the lower visibility of farmers as community members and collectively, as well as to their lesser representation on municipal councils. Moreover, farmers are contributing to decline in the agricultural centrality of their communes. Hence, the lack of a PLU (local zoning document)[18] in communes has encouraged speculation on unbuilt land, with landowners bringing pressure on local officials to allow the construction of buildings on their land. This has led to indiscriminate building. This type of building contributes to an inability to access some farm plots and to paths having inadequate access. Falling farmer numbers is reinforcing attitudes of professional individualism, with farmers making use of their dual status as working farmers and

---

17 Above and beyond fiscal considerations, this grouping was motivated by a desire to stand apart from Annecy, in order to maintain the rural (and agricultural) character of participating communes. When the first district newsletter came out in 1996, district president J.M. Nantua wrote, '... by creating this district with over 8 500 inhabitants, our goal was to reinforce the rural sector on the edges of the urban area' (*Intercommunal Bulletin*, 5, 2002, p.1). The general orientation was clear.

18 The PLU is the basic level (NUTS5) of the French planning system, which is now decentralized. It reflects land-management decisions made at the commune level.

landowners. Attitudes toward land conversion vary but with no agricultural commission, and with agriculture viewed in the territorial-planning commission as land management, effective control over conversion is left to communes.

## Conclusion

Peri-urban territories are sites of profound economic and social modification. The changes underway are diverse in nature and intensity, as well as being the product of a long history. Change sheds light on the double phenomenon that is fundamental to peri-urbanization, which is urban expansion and the positions adopted by rural areas subjected to urban influence. Economically there is often a desire to reinforce and put into motion a secondary economic centre or at least create checks and balances to the continuing expansion of an urban core. The size of the urban agglomeration in a city-region obviously plays a role in determining the intensity of peri-urbanization and the impact of resulting changes. These changes are confronted by the desires of inhabitants in impacted areas to construct their own identity and not become a land reserve for a nearby expanding city. This double phenomenon is observed in the study areas through the importance that is attached to the existence (or absence) of secondary centres within city-regions. Due to their own economic activity, secondary centres can promote a relatively autonomous form of development. Their internal dynamics intermingle with those of the urban core, which explains why peri-urbanization processes result from both complementary and (occasionally) divergent trends.

Increasing demand for more local services is also a major issue in these territories. Services are perceived as favourable factors in the quality of life offered, and their absence is a negative factor. What is more, above and beyond stores, schools and day-care centres, the environment and the quality of environmental management are seen as important 'services'. For example, waste management and water treatment represent major constraints on communes in that they are legally obliged to provide this service. That said, by creating these services, communes reinforce the quality of life for their citizens and pave the way for more housing units in their territories.

The many facetted notion of the quality of life also concerns agriculture, its future evolution and its current functions. In this field, the reduction in the number of farms is a constant and long-term trend in the study areas. This has elicited a number of different reactions. Certain communes actively favour agricultural activities as integral parts of their economic landscape, whereas others perceive agriculture as a means to maintain the territory and landscape, as a source of attractiveness for peri-urban areas.

This attractiveness helps draw city dwellers into peri-urban zones, which then raises concerns about social segregation. However, in-migrants are not homogeneous. They make housing relocation decisions based on different criteria (cost of land and transport, the quality of a site, landscape, etc.), which should lead us to view peri-urban areas as a mosaic of social classes, who are grouped according to income levels.

Yet in coming to rural areas, these in-migrants add to environmental stress, as urban-integration processes within city-regions are associated with job imbalances between central urban areas and the surrounding peri-urban communes. Over the long-term, there is a question over whether this imbalance will create a problem for the sustainability of development, for peri-urban zones are already confronted with increased commuting and a saturation of transportation infrastructure.

Faced with these strong changes, the positions adopted by peri-urban territories, which are very different in nature while being dependent on the intensity of links with the urban core of the city-region, reflect decisions made in terms of local development in these territories and in particular the manoeuvring room that elected officials manage to create in running their municipalities. Elected officials are major actors in constructing their territory and they attempt to establish its identity and consolidate its position with respect to neighbouring urban areas. If planners and regional developers have a leitmotiv on housing-employment equilibrium for a sustainable and coherent spatial development, this still questions the notion that sustainable partnership exists between rural and urban areas.

## References

Baumont, C. and Bourdon, F. (2002) Tendances récentes de la suburbanisation des emplois, *XXXVIIIme Colloque de l'ASRDLF*, Trois-Rivières 21-23 Août

Beaucire, F. and Saint-Gérand, Th. (2001) Les déplacements quotidiens, facteurs de différenciation socio-spatiales ? La réponse du périurbain en Ile de France, *Géocarrefour*, 76, 339-347

Bertrand, N. and Marcelpoil, E. (2001) L'environnement, support de l'autonomie des territoires périurbains, *Revue de Géographie de Lyon*, 76, 319-325

Bertrand, N. and Vollet, D. (2002) The role of territorial factors in the creation of service enterprises, *European Research in Regional Science*, 12, 223-241

Bertrand, N., Fleury, Ph., Perron, L., Tolron, J-J. and Janin, C. (2005) Politiques d'aménagement et multifonctionnalité agricole dans le sillon alpin, *Symposium PSDR INRA*, Lyon 9-11 Mars

Beuret, J.E. (1997) L'agriculture dans l'espace rural: quelles demandes pour quelles fonctions?, *Economie Rurale*, 242, 45-52

Beuret, J.E. and Kovacshazy, M.C. (2002) Aménagement et développement rural: au carrefour d'une demande qui s'affirme et d'une offre qui s'élabore lentement, *Ingénieries*, spécial, 15-23

Bidou-Zachariasen, C. (2004) Gentrification le tabou français, *Revue Esprit*, Mars-Avril, 62-64

Bourne, L.S. (1996) Reurbanization, uneven urban developement and the debate on new urban forms, *Urban Geography*, 17, 690-713

Briquel, V. (2001) L'avancée de la périrubanisation dans les Alpes du Nord françaises et ses liens avec la croissance de la population, *Revue de Géographie Alpine*, 89, 21-40

Briquel, V. and Collicard, J-J. (2002) *Introduction to a Method for Detecting and Classifying Peri-urban Areas*, NEWRUR Deliverable 2.2a–2.2b, Cemagref, Grenoble

Cairol, D. and Terrasson, D. (2002) Les aménités des espaces ruraux, un enjeu pour les politiques publiques, un enjeu pour la recherche, *Ingénieries*, n° spécial, 5-14

Cavailhes, J. and Schmitt, B. (2002) Les mobilités résidentielles entre villes et campagnes, in Ph. Perrier-cornet (dir.) *Repenser les campagnes*, de l'Aube, La Tour d'Aigues, 35-65

Cavailhès, J. and Selod, H. (2003) Ségrégation sociale et péri-urbanisation, *Recherche en économie et sociologie rurales*, Chapter 2, INRA 1-2/03, Décembre
Chalas, Y. (2000) *L'invention de la ville*, Anthropos, Paris
Chavouet, J.M. and Fanouillet, J-Ch. (2000) Forte extension des villes entre 1990 et 1999, *INSEE Première*, 707
Clément, P., Grillet-Aubert, A. and Guth, S. (2001) *Transport et architecture du territoire : Etat des lieux et perspectives de recherche*, PREDIT, DRAST, MELT, IPRAUS, Octobre
Commission of the European Communities (1999) *ESDP – European Spatial Development Perspective*, Office for Official Publications of the European Communities, Luxembourg
Donadieu, P. and Fleury, A. (2003) La construction contemporaine de la ville-campagne en Europe, *Revue de Géographie Alpine*, 4, 19-29
Donzelot, J. (2004) La ville à trois vitesses: relégation, périurbanisation, gentrification, *Revue Esprit*, Mars-Avril, 14-39
Dubois-Taine, G. and Chalas, Y. (1997) *La ville émergente*, de l'Aube, Paris
Fouchier, V. (2002) L'emploi à l'avant-garde de la rurbanisation, *Etudes Foncières*, 95 (Janvier-Février), 24-29
Giuliano G. and Small K.A. (1991) Subcenters in the Los Angeles region, *Regional Science and Urban Economics*, 21, 163-182
Goffette-Nagot, F. and Schmitt, B. (1999) Agglomeration economics and spatial configurations in rural areas, *Environment and Planning*, A31, 1239-1257
Gordon, P. and Richardson H.W. (1996) Beyond polycentricity: the dispersed metropolis, Los Angeles, 1970-1990, *Journal of the American Planning Association*, 62, 289-295
Goze, M. (2000) L'émergence d'une nouvelle organisation territoriale locale française, *Etudes et prospective, Territoires 2020*, DATAR, 2 (Décembre), 47-60
Granelle, J-J. (2002) Les marchés fonciers, causes ou conséquences de la ségrégation urbaine?, *Etudes Foncières*, 99 (Septembre-Octobre), 8-15
Harthorn, T. and Müller, P.O. (1989) Suburban downtowns and the transformation of metropolitan Atlanta's business landscape, *Urban Geography*, 10, 375-395
Hilal, M. and Schmitt B. (2003) Services aux populations, *INRA Recherche en économie et sociologie rurales*, Chapter 4, INRA 1-2/03, Décembre
INSEE/INRA (1998*) Les campagnes et leurs villes, Contours et caractères*, INSEE, Paris
INSEE Première (2000) *Forte extension des villes entre 1990 et 1999*, INSEE Première 707, Paris
Jaillet, M-Ch. (2004) L'espace périurbain: un univers pour les classes moyennes, *Revue Esprit*, Mars-Avril, 40-64
Krugman, P. (1996) Urban concentration: the role of increasing returns and transport costs, *International Regional Science Review*, 19, 5-30
Lacaze, J-P. (2002) L'étalement urbain hier et demain, *Etudes Foncières*, 96 (Mars-Avril), 7-9
Le Jeannic, Th. (1996) Une nouvelle approche de la ville, *Economie et Statistique*, 294-295, 25-45
Lecat, G. (2004) Documents locaux d'urbanisme et structuration des espaces périurbains, Paper at the Colloque de l'ASRDLF, 1-3 Septembre, Bruxelles
Livre Blanc du Lac d'Annecy (1978) Schéma de secteur du lac d'Annecy, *Livre Blanc*, Commission Mixte Locale du Schéma de Secteur du Bassin du lac d'Annecy, Annecy
LOADDT (1999) Loi d'Orientation pour l'aménagement et le développement durable du territoire du 25 juin 1999, *Journal Officiel de la République Française*, 29 Juin, 9515 sq
Manzagol, Cl. and Coffey, W.I. (2001) Redistribution de l'emploi et territoires métropolitains: la recomposition du périurbain en Amérique du Nord, *Géocarrefour*, 76, 349-357

Martin, S. (2001) Autonomie périurbaine, in E. Marcelpoil and A. Faure (coord.) *Cahiers de l'OIPRA, Journée d'étude sur l'autonomie des territories périurbains en Rhône-Alpes*, 13 Octobre, IEP Grenoble, 17-19

Martin, S., Bertrand, N. and Rousier, N. (2004) Régulation des conflits d'usage de l'espace agricole périurbain: l'importance des documents d'urbanisme et de l'échelle de résolution des conflits, *Journées d'étude Les conflits d'usage et de voisinage*, 11-12 Octobre 2004 Carré des Sciences CNRS-INRA, Paris

Massini, J-R. (2001) *Dictionnaire de l'intercommunalité*, SEFI

Michelangeli, L. (2002) *L'impact politique et juridique des typologies*, INSEE/INRA, Rapport Cemagref Janvier

Ministère de l'Equipement (1997) *Atlas des Paysages de Haute-Savoie*, CAUE de Haute-Savoie et DDE de Haute-Savoie, Ed. Villes et territoires, Annecy

Morand-Deviller, J. (2003) Droit de l'urbanisme, Mémentos Dalloz, Paris

OECD (1996) *Amenities for Rural Development: Policy Examples*, Organization for Economic Cooperation and Development, Paris

OECD (1999) *Cultiver les aménités rurales: une perspective de développement rural*, J. Beuret, Saika Y., Organization for Economic Cooperation and Development, Paris

Ogden, P.E. and Hall, R. (2000) Households, reurbanisation and the rise of living alone in the principal French cities, 1975-90, *Urban Studies*, 37, 367-390

Piron, O. (2005) Les dynamiques territoriales 1999-2003, première analyse des résultats de l'enquête censitaire, *Note de Travail*, Avril

Pumain, D. (1994) Villes et agglomérations urbaines, in J.P. Auray, A., Bailly, Ph., Derycke and J.M. Huriot (eds.) *Encyclopédie d'Economie Spatiale*, Economica, Paris, 111-125

Rambonilaza, M. (2002) Mise en œuvre de l'évaluation économique des aménités rurales en Europe: le cas des aménités environnementales, *Ingénieries*, n° spécial, 91-104

Rapey, H., Josien, E. and Servière, G. (2002) Entretien de l'espace par l'élevage, *Ingénieries*, n° spécial, 67-79

Rousier, N. (2001) Un développement économique périurbain à plusieurs niveaux, in E. Marcelpoil and A. Faure (coord.) *Cahiers de l'OIPRA, Journée d'étude sur l'autonomie des territoires périurbains en Rhône-Alpes*, 13 Octobre, IEP Grenoble, 30-31

SCEES (2002) 44% des exploitations dans l'urbain ou le périurbain, *Agreste Primeur*, 117 (Décembre)

Secchi, B. (2000) *Prima lezione di urbanistica*, Laterza, Rome

Tabard, N. (1993) Des quartiers pauvres aux banlieues aisées: une représentation sociale du territoire, *Economie et Statistique*, 270, 5-22

Tolron, J-J. (2002) L'agriculture périurbaine .... Un espace urbain pour des aménités rurales ?, *Ingénieries*, spécial, 81-90

Vanier, M. (1999) *Urbanisation et Emploi: Suburbains au Travail Autour de Lyon*, L'Harmattan, Paris

Vanier, M. (2001) L'intercommunalité périurbaine: inconnues et constantes, in E. Marcelpoil and A. Faure (coord.) *Cahiers de l'OIPRA, Journée d'étude sur l'autonomie des territoires périurbains en Rhône-Alpes*, 13 Octobre, IEP Grenoble, 14-15

Velz, P. (1993) Métropolisation et dynamiques d'organisation des firmes, Paper at the Colloque métropoles et aménagement du territoire, Université Paris Dauphine, 12-13 Mai, Paris

Chapter 5

# Urban Spread Effects and Rural Change in City Hinterlands: The Case of Two Andalusian Cities

Francisco Entrena[1]

**Introduction**

During the 1960s huge internal population movements took place across Spain, with about 4.5 million people changing places of residence (Romero and Albertos, 1993). The four Spanish regions that attracted the majority of these migrants were Madrid, Catalonia, Valencia and the Basque Country. The remaining Spanish regions saw population decline, which was especially intensive in Extremadura, Castilla-León, Castilla-La Mancha and Andalusia. From the second half of the 1980s, a noteworthy slowing in long distance migration occurred, in favour of intra-provincial movements. In parallel with this, even though a decline in the agrarian population that intensified in the 1960s persisted, there was a cessation in rural depopulation, which had characterized the 1960s and 1970s. As García Sanz (1994, 1997) has demonstrated, the decline in the agrarian workforce and change in the rural population were two independent processes in Spain. Hence, similar to other advanced economies, an inversion of previous migration trends had started by the second half of the 1980s, when many more people began to view rural areas as attractive places to live or spend their free time (e.g. Champion, 1989).

Significantly, the key factors that fuelled the 'rediscovering' of the 'qualities of rurality' did not emerge solely from a new understanding of rural space. In addition, this transformation was driven by a lack of affordable housing in many city cores and by a worsening quality of life in the same areas (alongside rising collective awareness of this). This led to a redefinition of rurality, which drew in new urban demands. This process continues today with more intensity than ever. One of the main consequences of this has been mounting urban spill-over into

---

[1] I should acknowledge the valuable contribution to this chapter of work carried out by the sociologists Maria del Río and Nieves Rodríguez. Under the direction of Francisco Entrena they undertook much of the fieldwork on which analysis in the chapter is based.

many rural areas, whose socio-economic and cultural circumstances have became increasingly connected to more or less proximate core cities for which they are hinterlands or have some type of functional dependence and/or exchange. Urban spill-over has been particularly boosted by a revaluation of rural space as a suitable space for living and spare time activities (e.g. Paniagua, 2002), by the possibility of finding cheaper land for building on city outskirts, by the growing material and social deterioration of urban centres, and by improvement in transport and communication infrastructure. All this has facilitated daily rural-urban interaction by numerous people for work and leisure, as well as for personal relationships and contacts between inhabitants in relatively distant places. What this chapter seeks to understand, in the context of two case study areas in Andalusia, Spain, is how the changing relationship between rural and urban areas has impacted upon city hinterlands by provoking their socio-economic and functional restructuring, at the same time as bringing about a cultural redefinition of the significance of these territories.

**Urban Spill-over and its Effects on Rural Areas in Spain**

In spite of its growing importance, the Spanish literature on the phenomenon of urban spill-over is scarce. Contributions to this literature tend to emphasize different aspects of the dispersion of urban influences. For instance, Zárate (1984, pp.102-103) highlights how, what he typifies as the 'rururban fringe', is a social space characterized increasingly by urban styles of life, by greater population mobility, more social variety, differentiated social behaviour and a contemplative valuation of nature. From another perspective, Pellicer Corellano (1998) underscores the 'peri-urban nature' of many territories affected by urban sprawl, given that these territories constitute spaces that are in conflict with or at the interface between rural and urban sub-systems. For this reason, it is argued, it is in peri-urban areas that greater tensions between resources are concentrated. Thus, if 'natural' space or a fruit-and-vegetable area are by the door of a city, this adds positional value, as well as generating conflict, which is of greater intensity than if these areas were located many kilometres from the city. The physical dimension of conflict may also be appreciated by studying road and energy networks, which intertwine the city and its periphery, or in social fallows or spaces awaiting new uses. These areas are often degraded ecologically, as well as being subject to legal redefinitions of their function, so they are susceptible to speculative interests. Given a deterioration of many peri-urban spaces in Spain in recent decades, which, above all, has resulted from intense and uncontrolled processes of urban sprawl, Pellicer Corellano (1998) points to the need to re-establish equilibrated relations between the city and its environment. For Pellicer Corellano, this requires the restoration of degraded peri-urban spaces, so as to bring about ecological, cultural and economic revitalization, with the aim of satisfying new social demands for an improved quality of life.

Highlighting the depth of sentiment favouring change, Sancho-Martí (1989) has argued that, as a consequence of urban dynamics, technological innovation and the motivations of their users, city hinterlands are now in continual crisis, which means their functions are in a state of persistent redefinition. This is not simply because of new demands on existing resources, but also arises because these areas receive functions that are expelled from the heart of the city.

When studying city hinterlands, Spanish commentators lay heavy stress on their diffuse nature and borders. By emphasizing this point, they allude to the fact that within these areas are types of work and ways of living that reveal an intermediate socio-economic and spatial position between the rural and the urban. In this regard, Precedo Ledo (1996) refers to peri-urban areas as 'urban regions', which are shaped by a preponderantly rural space in which those who work in the city live. Yet usage of the expression 'urban region' is woolly. Sometimes it means a regional space organized by a city, as with the concept of a nodal or polarized region. At other times the expression denominates a large metropolitan area, which in a few cases is interpreted as an urbanized region. This last interpretation, according to Precedo Ledo (1988, p.90), refers to a dense, complex, inter-urban structure that is constituted by a network of scattered urban settlements, the urban nebula and associated rururban areas, which has urban socio-economic characteristics and is functionally constituted as unitary space. This settlement structure represents one end of a spectrum of rural-urban relationships, in which the more common form city hinterlands take see gradations and diversity of urban and rural interaction, with some rural zones having relatively weak ties with their nearest city, while others are intimately bound into its cultural, economic and social fabric.

Amongst this settlement structure, small town and rural areas in the peri-urban or rururban zone are integrated parts of urban or urbanized regions, whose 'diffuse nature' not only refers to the undefined character of their geographic or physical boundaries, but also to their socio-economic characteristics. In other words, the borders of city hinterlands are unclear, not only because it is physically and geographically difficult to establish accurately a division between a city and its hinterland, or between one city-region and another (they commonly overlap; Coombes, 2000), but also because these areas demonstrate socio-economic characteristics that are in the process of change and redefinition. This is a key feature, not only of rural areas with strong urban linkages, but of all those country spaces, even those relatively far from a city, which are somehow affected by its influence, so they constitute parts of the city's hinterland. Certainly, urban spill-over effects in 'secondary' urban spheres of influence are more blurred, since there is often no sizeable or clearly visible enlargement in physical growth, but a spread of typically urban socio-economic characteristics, which lead to change in cultural habits, ways of life, standards of household facilities or kinds of work, while traditional agricultural activity declines gradually. Indeed, in-migrant urbanites help bring such changes into being, with demographic pressure boosting mutations. As a consequence, even though municipal and national databases might reveal no major increase in inhabitants, the fact that peri-urban areas can be visited daily by a noteworthy volume of urban residents (especially during holiday periods or at weekends) increases urban impacts considerably (e.g. Mansvelt Beck, 1988;

Hoggart and Paniagua, 2001). An example of these 'blurred' urban effects is analyzed below in the case of La Alpujarra.

Importantly, irrespective of the tightness of urban linkages, within city hinterlands it is possible to observe a sort of hybridization between the rural and the urban. The term 'hybridization' alludes to the fact that such areas maintain many of the typical attributes of rural ambits, at the same time as they are socio-economically linked to a core city, which in turn enhances the chances for multi-functionality amongst inhabitants. In truth, such multi-functionality is nowadays characteristic of the majority of rural areas (e.g. Huylenbroeck and Durand, 2003), as they are all submitted in some way to urban influence. The Spanish literature echoes this fact.[2] This means it is no longer adequate to conceptualize rural-urban dichotomies, as analysts commonly used to in the past (Sjoberg, 1964). Contrary to these dichotomies, current Spanish policies and commentaries on rural areas assume, at least theoretically, the need to achieve balanced, complementary and co-operative urban-rural relationships. In this regard, current Spanish thinking is in accordance with European orientations toward achieving balanced socio-territorial development, as presented in the European Spatial Development Perspective (ESDP) (Commission of the European Communities, 1999; Crecente *et al.*, 2001).

At this point two ineluctable key questions arise. The first is, how can we measure the effects of the spread of urban influence over hinterland territories? The second is, to what extent do these effects bring improvements in the socio-economic dynamism of such territories? A supplementary to the second question is, how far Spanish development practices effectively comply with the ESDP aim of achieving balanced socio-territorial development? In order to answer these questions, two case studies are subjected to in-depth analysis. Before moving onto this analysis, the configuration and general characteristics of Andalusia's urban system are briefly summarized in order to contextualize the case studies.

**Paradigmatic Relevance of the Andalusian Case**

The socio-economic transformation and demographic movement that were experienced in the second half of the twentieth century in Andalusia made this southern European region a highly suitable exemplar for understanding the 'spread' of urban influences in Spain. Similar to other parts of Spain, Andalusia in the 1950s was a rural society, where small towns prevailed. Its eight provincial capitals were then mainly devoted to administrative services for their respective territories, and, as with some other cities, acted as centres for the transformation and commercialization of agrarian production within their areas of influence.

This situation was modified due to the modernization and industrialization of the Spanish economy in the 1960s (Salmon, 1995), when there was a massive shift

---

2   In this context, note the title given by the organizers to the IV Spanish Portuguese Colloquia for Rural Studies, which took place in 2001 in Santiago de Compostela. This was entitled 'The Multi-Functionality of Rural Spaces in the Iberian Peninsula'.

of rural Andalusian inhabitants toward more industrialized areas. This emigration brought gradual demographic decline to most rural municipalities. In parallel with rural outmigration, many of the main Andalusian cities began to experience population growth (Marchena Gomez, 1984). Such processes became particularly strong after the second half of the 1980s, due to a natural demographic increase in the cities, continuing rural population losses and the return of many Andalusian migrants from abroad and from other parts of Spain.

A general worsening in the social situation was brought about by worldwide economic recession in the 1980s and 1990s, but the arrival of democracy in Spain, with the ensuing establishment of an autonomous government for Andalusia, reinforced the tendency for the return of migrants. This was because regional autonomy brought advances in health, education and welfare services, so that, while it remains one of most backward regions of the European Union, Andalusia has experienced noticeable progress in living standards. The regional government has made noteworthy improvements in infrastructure, which have contributed significantly to strengthening the incipient course of economic growth, which began in earlier decades. Without a doubt, demographic growth associated with economic progress has increased the spread of urban tentacles over the countryside, even if urban influence varies in its intensity.

As an outcome of all the aforesaid, a new urban system has been fashioned, with population concentrated in provincial capitals, and in intermediate and larger cities along the Guadalquivir river valley and, above all, on the coastal fringe. In the mountainous interior there is demographic 'emptiness'. As shown in Figure 5.1, the largest group of proximate municipalities that have seen recent substantial growth are primarily in densely inhabited areas on the coastal fringe and in the Guadalquivir Valley, while inland areas to the north have seen little growth or even decline. Regarding provincial capitals, population increases were recorded for Almería (10.2 per cent), Jaén (7.9 per cent), Córdoba (3.9 per cent), Sevilla (2.9 per cent) and Málaga (2.3 per cent). In these cases, processes of urban spread occurred in parallel with the 1990s socio-economic development of each capital. However, Cádiz (10.6 per cent), Granada (4.7 per cent) and Huelva (0.9 per cent) experienced population losses. Decline is especially evident in Cádiz, where losses in the centre have not been compensated by growth on the urban periphery, which is a result of a general 1990s crisis in the shipyard industry in the towns of the Cádiz Bay. By contrast, the relative socio-economic strength of Huelva has made significant demographic increase in surrounding municipalities possible. These have counterbalanced demographic slippage in the city core. A similar trend has happened in Granada, but in this case with more intensity, so the deconcentration processes that started in 1991 have broadened, especially into the city's hinterland.

**The Case Study Areas**

Municipalities surrounding Granada City saw significant demographic increases in the last inter-census decade, which is one reason for selecting this city-region to explore the impact of urban spread influences on city hinterlands. Moreover,

Granada is an archetypal case for studying urban development, as its driving force is tourism, which has been so important in the take-off of Andalusia's economy over the last 40 years, as well as in shaping regional spatial planning (Fernández, 2004). While the major impacts of tourism are in shoreline areas, this derives from mass tourism, for which urban sprawl and widening urban influences are not explained by interactions between cities and their hinterlands. Rather, coastal dynamism is the outcome of a worldwide diffusion of an image that the Andalusian coast (especially the so-called 'Sun Coast') is an attractive venue for leisure. This is why, regarding the littoral, the area of El Ejido (a large, new municipality 36km west of Almería City) has been selected as a second case study area, since urban pressure on rural areas has risen sharply in this area, fuelled by the rapid spread of intensive fruit and vegetable greenhouse farming. As will be shown later, farm growth in El Ejido has been accompanied by tourism development, above all in the last few years.

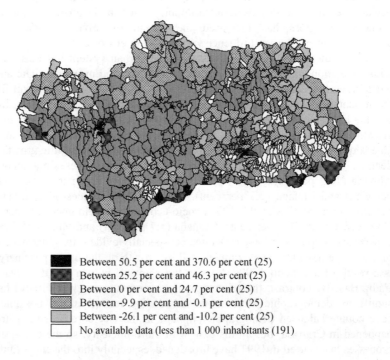

Between 50.5 per cent and 370.6 per cent (25)
Between 25.2 per cent and 46.3 per cent (25)
Between 0 per cent and 24.7 per cent (25)
Between -9.9 per cent and -0.1 per cent (25)
Between -26.1 per cent and -10.2 per cent (25)
No available data (less than 1 000 inhabitants (191)

**Figure 5.1 Percentage Population Change in Andalusian Municipalities 1991-2001**

Source: Computed from statistical data from the bank 'La Caixa'.
Note: Beside each caption, the figure in parentheses is to the number of municipalities.

The intention in studying two city-regions with quite different economic bases, was to examine how far the spread of urban influences on rural zones yields similar consequences and changes, independently of the driving force that causes them. In each city-region, three areas were subjected to detailed investigation. In El Ejido, these municipalities were Berja, Adra and Albuñol (respectively, located at 21, 22 and 41km from El Ejido City). In Granada, the municipalities of Monachil and Güéjar Sierra (at 8 and 16km from the City) and part of the extensive zone of La Alpujarra, which is about 60km south of City, were selected. The La Alpujarra zone encompasses territory in the municipalities of Lanjarón, Bubión, Capileira, Pampaneira and La Taha.

A key reason for choosing these locations was to ensure coverage of places with dissimilar intensities of urban linkage. Another factor in the selection was to explore how urban influences do not come solely from proximate urban centres, but also from more global forces. As shown below, factors of a more 'global' nature are critical to the huge development that has occurred in El Ejido and also help explain key changes in Monachil and La Alpujarra.

For the purposes of this investigation, the analytical framework that has been adopted is to explore relationships between forces that drive change in an area and the nature changes exhibited. Examining through the lens of three geographical zones, the pages that follow provide an exploration of change outcomes where the main factors fuelling transformations are different. Thus, for Monachil, the development of the ski resort of Sierra Nevada and the growth of commuting to Granada City have been instrumental in inducing rapid change. For Güejar Sierra and La Alpujarra commuting to Granada is relatively inconsequential but the Natural Park of Sierra Nevada and rural tourism have produced a noteworthy socio-economic dynamism, which is particularly intense in La Alpujarra. Then, in El Ejido, the drivers of change expansion of modern intensive agricultural and glocalized urban spread effects.

**The Ski Resort and City-Ward Commuting in Monachil**

Monachil is a very suitable example for showing how the spread of urban influence and the ensuing changes this brings can be exacerbated when commuting growth, which is usually the key factor in the spread of a city's tentacles over the countryside, acts in conjunction with another factor; which in the case of Monachil is the development of a ski resort. This resort, which is named Pradollano, is located in the mountains of the Sierra Nevada, which is internationally known and highly valued as a tourism zone, especially after major improvements were made to the resort's facilities when it hosted a worldwide competition for Alpine Skiing in 1996. Both commuting and the ski resort have decisively impacted on Monachil, which has experienced amongst the most intense urban pressure in the Granada commuter belt. This is reflected in the municipality seeing significant demographic growth during preceding decades. Today, the municipality of Monachil has three urban nuclei; namely, the traditional core, which is known as Monachil-pueblo, the so-called Barrio de la Vega, which is a recent development, and the ski resort of Pradollano, which has experienced a notable enlargement over the last few years.

In addition to commuting and ski-related development, expansion in Monachil has been encouraged by the superior quality of its natural environment and landscape. Being located within Granada's hinterland, these factors have helped make the municipality an increasingly attractive residential choice and tourism destination. As a result, since 1970, approximately 1 000 new homes have been built in the municipality, including second homes. However, as Table 5.1 shows, this growth has been most marked for primary residences, which increased by 56.0 per cent between 1991 and 2001. That this growth has been associated with social change is further suggested by a lessening in the density of household occupation, with the municipality's population having risen to 5 684 in 2003, which represents a 34.8 per cent rise on 1991.

Table 5.1  Number of Dwellings in Monachil by Occupancy, 1986-2001

|  | 1986 | 1991 | 2001 |
|---|---|---|---|
| First homes | 874 | 1 092 | 1 703 |
| Second homes | 1 598 | 2 113 | 1 179 |

Source:   Institute of Andalusian Statistics.

These changes have occurred in parallel with gradual decline in agriculture, which has been on a downward slope since the 1960s. Associated with this fall, the local manufacturing sector, which, in the form of food processing and exploitation of other local natural resources, suffered a deep crisis as a consequence of a failure to invest in modern machinery and the development of more intensive methods of production elsewhere. This led to the disappearance of most hydroelectric plants, flour and olive oil mills and tile factories.

Table 5.2  Percentage of the Monachil Workforce in the Construction Sector, 1950-2001

| Year | 1950 | 1960 | 1970 | 1986 | 1991 | 2001 |
|---|---|---|---|---|---|---|
|  | 2.2 | 7.4 | 20.2 | 18.1 | 19.6 | 18.8 |

Source:   Institute of Andalusian Statistics.
Note:       The peak in 1970 is because this was the period in which Barrio the La Vega was constructed.

The crisis of the traditional economy forced many people to emigrate in the 1960s. However, some years after, the majority of migrants returned, although on return they tended to settle in the Barrio de la Vega (rather than in Monachil-pueblo), where they are employed in the building and service sectors.

These sectors were able to absorb these workers owing to their growing importance within the municipal economy. Table 5.2 reveals this by showing the evolution of the working population that was employed in the construction sector. To a large extent, this expansion is a consequence of the economic and demographic dynamism experienced by Granada City, which propelled expansion well beyond its older quarters, bringing the construction of new neighbourhoods in its hinterland.

The spread of urban influence went hand-in-hand with rising diversification and complexity in local socio-economic and occupational structures. Service industries experienced significant growth in total employment, as can be seen in Table 5.3. Demand for labour from the ski resort of Pradollano was also decisive in pushing expansion forward. As a result of socio-economic change, each of the three urban nuclei in the municipality now specializes in a different economic sector. Monachil-pueblo is characterized by agricultural production, Barrio de la Vega residents find most of employment in construction or service sectors, with many commuting for work, especially to Granada City, while Pradollano is dominated by tourism.

**Table 5.3  Percentage of the Monachil Workforce in Service Industries, 1950-2001**

| Years | 1950 | 1960 | 1970 | 1986 | 1991 | 2001 |
|---|---|---|---|---|---|---|
| Percentages | 11.2 | 13.3 | 16.5 | 46.4 | 62.5 | 69.5 |

Source:   Institute of Andalusian Statistics.

The aforementioned changes brought noticeable improvements in welfare indicators, at the same time as the municipality experienced the emergence and a continual increase in demand for new services to cater for the growing population. This contributed to the creation of a large number of new facilities and job opportunities, as well as a noteworthy rise in the wealth of the municipality. In addition to new services for the resident population, in Pradollano a noticeable increase in services was aimed at tourists and skiers. These new service types brought some benefits for the local residents, who either enjoy them directly or have gained from increased opportunities for employment. In this context, the municipal council is fully aware of the need to improve and expand services, since both improve standards of life for municipal residents, as well as helping make the area more attractive for visitors, which further enhances the prospects of local economic gain. In this regard, since 1991, health facilities, waste collection, water and sewer systems, plus safety and security, have all been provided and/or notably improved in each of the three urban nuclei. The main source for financing these improvements has been the municipal administration in Monachil-pueblo and Barrio de la Vega, but in Pradollano decisions on building licenses and further urban development were largely in the hands of the national and the Andalusian governments.

Significant progress in service provision was recognized by almost all interviewees during fieldwork.³ Services are now generally seen to be at a higher level than in most municipalities of similar size and character. As one spatial planning regulator expressed it:

> A strange problem is that we have a census, which has about 6 100 inhabitants, but we are giving services to one and a half million of people, as if this is a great city, due to Sierra Nevada. Many services are also given to Granada City's inhabitants who come at weekends to visit us. The growth of the services has been unbelievable.

As a manager of a hotel complex expressed it:

> The town has changed a lot in terms of its services. There are now services like those of a large town. Before there was not good access to the town by road, now there is very good communication. At the moment we have all the improvements that a much bigger town can get: institutes, schools, library, culture house with many services, quite a great infrastructure at all the levels.

The main service gains have been in education, health, culture and transport. Commercial services growth has been less noteworthy, mainly because people shop in the City. Many local people not only go to the capital for work, but also shop there and use Granada for health, culture and leisure services. In this Monachil is not exceptional, but a paradigm for the metropolitan region. In the whole of the city's hinterland, employment has passed from being essentially agricultural to employment based, to a great extent, on the construction and service sectors; which has been especially noteworthy in terms of growth in new services in education, health, culture, transport and, above all, in the spread of tourist activities. As a result, economic activity rates have risen significantly in the commuter belt, although they are still inferior to those of the City.

This pattern requires some clarification. It is true that most people who move to live in commuter belt territories already have a job. They rarely leave the City and look for employment locally. Some move to the commuter belt for the purpose of setting up a company or because they have a job contract in the area. But the majority move into the hinterland with a job already. If they are unemployed or become unemployed, then they prefer to register for work possibilities in Granada City, where there are more numerical job opportunities, as well as a greater variety of work openings. This means they are entered as official residents of the commuter belt but as unemployed workers in the City. As one official in the City Council's Development Department informed us:

> We import jobless people because there are more job offers in the capital. There are more possibilities to find employment here and people from the metropolitan area register here with the INEM [National Institute of Employment]. These are people

---

3   The fieldwork on which this chapter is based included 108 in-depth interviews, which were conducted between July 2002 and January 2004. These interviews were with persons in a variety of different roles concerned with changes in the Granada and El Ejido hinterlands. Interviewees had expertise in the fields of residential expansion, employment generation, household services, consumer groups, policy formulation for spatial planning, and policy regulation and/or implementation.

who are not included in the population census of the capital, but they appear as unemployed persons in the databases of the capital.

For a great number of individuals, rural commuter-belt municipalities are selected solely for residential purposes, not for work. Employers in the hinterland are well aware of this, with one summarizing the situation as:

> Many people who come to live here arrive with a job. They do not come to look for a job here. There are many more people working in the capital and living in the 'town' than the opposite.

The resulting absence of workplaces near places of residence, with a consequent need for daily commuting, is recognized as problematical, with the logic of people's concerns captured by one agent involved with employment generation:

> Geographical mobility is a complex issue, as individuals have not prepared themselves mentally for it but they have to do it, because industrial parks are not located in the City. There are many jobs to be found, but they are not nearby. Many people do not have a driver's licence and this makes the fact of daily commuting to relatively distant employment more difficult for them.

In short, employment opportunities in Granada City are much greater than in its commuter-belt, which is a key reason for high levels of daily commuting between the metropolitan hinterland and the City (as well as between some towns in the hinterland).[4] Commuting provides opportunities to do other things in the City, such as shopping and a diversity of leisure activities, which means many residents of the rural hinterland use services and facilities in the capital, which is where they spend a great part of their lives. It also means that commitments to a municipality of residence are largely restricted to sleeping and weekends. This leads frequently to weaker social (and service) benefits for these municipalities than their population growth might lead one to expect, as well as challenging core values in the European Commission's European Spatial Development Perspective, as the settlement structure encourages greater daily mobility as well as dispersed forms of urban expansion (Commission of the European Communities, 1999).

## Tourism in Güejar Sierra and La Alpujarra

As mentioned above, rural tourism is a key driver that helps explain important changes in Monachil. However, it is in Güejar Sierra and, especially, La Alpujarra where rural tourism and nature tourism have been major causes of transformation in recent decades. In the case of Güejar Sierra there is an incipient rural tourism, which is favoured by its nearness (16km) to Granada City. This helps attract

---

4  One indication of the dominance of the capital, the 2001 index of local economic activity computed by the Bank 'La Caixa' scores Granada at 614, with the next largest value in the metropolitan region at 32 (Monachil is ranked 11th with a score of 12). This index is based on tax values for industrial, commercial and service activities. The index expresses participation of municipalities in national economic activity (100 000 is the score for all Spain). The index is a good indicator of economic activity levels.

visitors and trippers searching for good natural and landscape characteristics in the mountainous territory of the municipality, which is located in the Natural Park of Sierra Nevada. However, Güejar Sierra has seen change despite relatively scanty influences over the area from the core city of Granada.

By contrast, while La Alpujarra is located in the Natural Park, it is on the southern side of Sierra Nevada. This means that it is relatively distant from Granada City (about 60km), with weaker city-ward commuting pressures. Unlike Güejar Sierra, in La Alpujarra rural tourism has been favoured by the worldwide spread of an image of the Sierra Nevada as a paradigm for isolated places with peculiar cultural traditions and ways of life, which, at the same time, have attractive natural environments and landscapes. The fact that urban influences are greater in La Alpujarra than in Güejar Sierra demonstrates that proximity to a city core is not always decisive in urban spread effects. These can depend upon many factors, one of which is the growing global diffusion of the supposed qualities of particular types of territory, which subsequently increase the attractiveness of some places (like La Alpujarra) for increasing numbers of visitors and in-migrants.

This linkage with the global is often a more powerful cause of change, as in La Alpujarra, than city 'spill-over' effects, as demonstrated by the relatively stagnant situation of Güejar Sierra. In fact, in this municipality there has been relatively little new residential development in comparison with Monachil and, indeed, in contrast with La Alpujarra. Thus, Güejar Sierra has seen a decrease in the number of first homes over time, which is in line with demographic change in the area, as the population has diminished, at the same time as experiencing a gradual ageing. Yet there has also been growth in second homes, which clearly shows reinforcement of linkages with the Granada. Yet this reinforcement has not led to important changes and/or improvements in facilities for the local population. The municipality has centres for primary, secondary and adult education, but these only satisfy basic education needs, and there is no public library, nor are there sufficient sports and recreational facilities to satisfy the current population's needs. Indeed, limited development in the municipality is revealed in traditional subsistence agriculture still being a noteworthy economic activity, although farm income is a complementary source of funds for the majority of families, many of whom have passed from working in the primary sector to employment in construction, and, to lesser degree, in tourism. Commuting levels are relatively low, being most notable amongst women who travel daily to Granada to work as domestic servants (there are 10 bus services to Granada a day, with 12 from Granada). All this means that the unemployment rate in Güejar Sierra is higher than the average for Granada's hinterland, primarily because new job formation is weak. As one informant put it:

> In fact a proportion of employment has been destroyed due to the abandonment of agriculture and cattle breeding. Maybe some employment has been created in the hotel sector. The people have started to move into rural tourism, gastronomy... In this sector some new occupations have been generated, but in others they have been destroyed.

There are, however, promising signs that potentially profound changes could bring significant development to the still embryonic sector of rural tourism. In this regard, workers leaving agriculture occurs hand-in-hand with a gradual increase in tourism activities, which have in their favour the appeal of the mountainous territory of Güejar Sierra. This area is very suitable for sport and other nature-centred leisure activities. An additional factor strengthening this trend is the fact that the service sector is oriented toward tourism, for the municipality possesses accommodation and catering outlets that are adequate for current tourist volumes. For sure, this represents an inferior level of provision to that of La Alpujarra and, above all, Monachil, but it does provide a suitable base for future development.

In spite of the shortcomings they suffer, people in Güejar Sierra have a generally positive view toward the services at their disposal. In truth this probably owes less to the suitability or sufficiency of such services, and it does not reflect sufficiency for expected future needs if rural tourism continues developing. Rather it owes more to the fact that the past was noticeably worse than today. Even so, there are symptoms of dissatisfaction, related to the consequences of what are seen as increasing urban influences on the area:

> Güejar Sierra and the surrounding towns have always tended to improve. If you compare them with the situation 30 years ago the change has been quite positive. Maybe some aspects like leisure, sport and health have been neglected. In the past the doctor was here all year round, but with [service] restructuring you do not have a year-round doctor now. Regarding education, something similar has occurred. Before the teachers lived in the town, [but] now it is different. The teacher does his work and then disappears. When the doctors and the teachers were here, the contact with the population was better than now. (Provincial Government Official)

> People are very unhappy about the school as 14-year-old children have to leave, to go to Granada, and it is a very difficult age for doing this. Children have to take the bus very early, at 7 in the morning, and they return very late, at 4 in the afternoon. (Local Councillor)

In terms of the municipal capacity to respond to its service needs, there is a lack of finance for the local administration, which leads to low levels of service provision. In addition, proximity to Granada City and private car use encourage commuting to the City. This is important in the deterioration of educational and health services. Yet proximity has positive effects, such as improved transport services and more tourists, which offer the main potential for development in this municipality.

In contrast to the limited development that has been experienced to date in Güejar Sierra, rural tourism is a key factor in the transformation of La Alpujarra. In this area we have witnessed continual increases in housing from the 1970s. In this regard, between 1994 and 2000, except in Lanjarón, which is the largest municipality of those studied, the number of second homes rose in all the municipalities analyzed from La Alpujarra. This suggests that, in the last few years, the construction sector in these municipalities has been mainly oriented toward the tourism sector.

**Table 5.4  Percentage of Workforce by Economic Sector in La Alpujarra Municipalities 1981-2001**

| Municipality | Agricultural sector | | | Manufacturing sector | | | Construction sector | | | Service sector | | |
|---|---|---|---|---|---|---|---|---|---|---|---|---|
| | 1981 | 1991 | 2001 | 1981 | 1991 | 2001 | 1981 | 1991 | 2001 | 1981 | 1991 | 2001 |
| Lanjarón | 32.6 | 25.8 | 10.3 | 19.3 | 16.6 | 14.1 | 12.6 | 13.2 | 16.1 | 35.4 | 44.4 | 59.5 |
| Bubión | 42.1 | 7.6 | 0.8 | 15.8 | 8.6 | 8.9 | 21.1 | 15.2 | 12.9 | 21.0 | 68.6 | 77.4 |
| Capileira | 71.7 | 46.9 | 13.3 | 13.0 | 8.6 | 8.3 | ... | 10.9 | 17.8 | 15.2 | 33.6 | 60.6 |
| Pampaneira | 61.8 | 13.8 | 4.3 | 11.7 | 24.1 | 7.5 | 2.9 | 25.9 | 17.2 | 23.5 | 36.2 | 71.0 |
| La Tahá | 83.8 | 35.4 | 5.8 | ... | 13.3 | 17.5 | ... | 10.0 | 13.8 | 16.2 | 41.3 | 63.0 |

Source: Computed from data in the 1981 Andalusia Census and Institute of Andalusian Statistics data for 1991 and 2001.

Similar to Monachil, rural exodus was suffered in La Alpujarra from the 1960s, with depopulation one result. Nowadays this process has been reversed, as a result of the spread of rural tourism. La Alpujarra municipalities affected by tourism have undergone noticeable economic change, as shown by adjustments in economic activity over time, which reveal sharp shifts away from the primary sector toward tertiary sectors (Table 5.4). But the decline in agricultural employment has not only been compensated by increased service sector activity but also, to a lesser extent, by employment growth in manufacturing and construction. Even so, the spread of rural tourism has been the key factor in bringing about change. In Lanjarón, the local spa has been a traditional tourist attraction (and the area has mineral water production). By contrast in the so-called Barranco del Poqueira (where the municipalities of Pampaneira, Bubión and Capileira are located), the main attraction for rural tourism has been the preservation of cultural traditions and vernacular architecture, which have combined to project an imagery that is typical of secularly isolated small, mountainous villages.

Significantly, tourism has been important in curbing unemployment increases, which resulted from agricultural decline that began in the 1960s. The labour force is now mainly concentrated in services and in construction, with the expansion of both closely related to the spread of tourism. However, jobs in these sectors tend to be short-term, which means that the unemployment rate goes up and down seasonally. Even so, demand from the growth of rural tourism is leading to more regularity of employment. Indeed, as one commentator on development in this area informed us, demand from rural tourism has now reached the point where there are some bottlenecks in service capacities:

> La Alpujarra needs more workers. Tourist opportunities are [now] greater than the supply of workers. People use their rural homes for tourism, charging rents [for vacation lets], or they live off unemployment subsidies, but it is difficult to find workers. For example, if you need a plumber you must wait for a long time ...

A consequence of this is the in-migration of workers, the majority of which remain in the area only during peak periods. The seasonal arrival of migrants, along with the long-term establishment of households from elsewhere in Spain or from outside Spain, contribute to an undermining of local identity. In some cases, this is provoking small brushes with traditional residents, owing to different understandings of how new developments are affecting the area.

One of the outstanding effects of tourism has been a noteworthy improvement in services for the local population, with subsequent progress in living conditions. From being a paradigm of a traditionally agricultural society in decline, La Alpujarra has been transformed into a rural society that is increasingly tertiarized, whose inhabitants have more opportunities for paid-work as tourism expands. However, in spite of evident advances in education and health services, there are still sizeable deficits, such as the fact that cultural and sport facilities are lacking and usually depend on schools. Moreover, public transport in La Alpujarra is deficient. So, except for Lanjarón, where there are nine daily bus connections with Granada in both directions, the other municipalities have only three connections. In addition, journey times are lengthy, as settlements are linked by winding roads. This explains complaints from interviewees regarding services at their disposal:

> There is a deficit. The municipalities are overpopulated in the seasons of Christmas, Holy Week and during summer, as they do not have enough services.

> Improvements have been very few. We are a little 'abandoned'. The doctors come here three days in the week, for two hours in a morning.

> When the children finish basic school, they have to go out of town for a secondary school. There are children that have uncles, cousins or brothers living in Granada, and so some of them move to live in the core city for studying. Others commute each day for this purpose to Granada or to the nearest town of Órgiva.

Unlike these views, which are from local commercial, non-governmental and voluntary sectors, politicians have more positive opinions about services:

> The schools are very good. We are building some new schools adapted to the LOGSE [the last reform of the 1990s for secondary education]. They will have heating, which we [when we attended school] did not have.

> We have a small clinic and a doctor. He does not come as often as we want, but does come every day except Wednesday.

> In Bubión there was a sport centre with two broken goals. Now we have a good sport centre and a gym. We also have a cultural centre with one person dedicated to it. This [the services situation] has changed a lot. In Bubión you can find some computer courses subsidized by the town council, courses of gymnastics, yoga, maintenance, English ... and a lot of people from Capileira come to use them.

One reason for shortcomings in services is a chronic deficit in the organization of the local administration, as there is insufficient revenue to finance services, above all because seasonal tourism brings a temporary overpopulation, comprised of those who do not pay local taxes but demand more and more services.

At this point, a key challenge for present trends in La Alpujarra and emergent transformations in Güejar Sierra is that national and regional authorities become aware of the need to compensate municipal fiscal deficits associated with the provision of services. This is because services still need notable improvement, given that the strengthening and continuity of the present gains brought about by rural tourism, with an ensuing advancement in living conditions, both closely depend upon such improvement.

## Intensive Agricultural Production and Glocalized Urban Impacts in El Ejido

El Ejido has undergone intense urban development in recent decades. This change is revealed in the fact that it has passed, from the beginning of the 1960s, when it was an area of disperse housing belonging administratively to the town of Dalías (up to 1982), to an urban centre that today has a population of more than 50 000 inhabitants. This major change cannot be understood without reference to profound transformations in the agrarian sector. Thus, at the start of the 1960s, El Ejido was a small centre, in an underdeveloped, disperse setting, that was located in a rural area of poor scenery, which was devoted to traditional agriculture, even though for the most part the land was uncultivated and used by cattle. In contrast, nowadays it is not possible to distinguish between El Ejido City from its former surroundings, which are swarming with a multitude of greenhouse installations and numerous paths for accessing to them. The City's surroundings have been transformed completely, into a scenery referred to by many local people as the 'plastic sea', on account of the characteristic immense plain of plastic covered greenhouses. This peculiar greenhouse landscape is not only a feature of the area adjacent to the El Ejido core, but extends across rural territories in municipalities that are relatively close to this core, and consequently are more or less heavily impacted by changes in the core. This is the case of the municipalities of Adra, Berja and Albuñol. As a result, all these territories have experienced a restructuring of their socio-economic role and their cultural significance has shifted in a manner that is usual when societies evolve from traditional agricultural ways of life to modern ones. In El Ejido city-region, restructuring has been especially intense.

Deep transformations have occurred in agriculture in the whole of Almeria, the Andalusian province in which El Ejido is located. In this province, similar to many other parts of Spain, during the 1960s agriculture was confronted with crisis. A large surplus existed in the agrarian workforce, which was brought about by decline in traditional agriculture, which started in that decade. These surplus workers could not be absorbed by other economic sectors, as there was little manufacturing or tourism, and services were undeveloped. These circumstances began to change in the 1970s, which were a crucial time, when restructuring within the agricultural sector led to economic take-off in Almería Province. This take-off, which is the chief factor explaining transformation in the city-region of El Ejido, was boosted by the introduction of intensive agricultural production methods, based on the greenhouse cultivation of fruit and vegetables for international export (Tout, 1990), mainly to the European Union.

The fact that urban growth in El Ejido has been deeply linked to the development of modern agriculture, which focuses on international exports, alongside the manner in which these changes have attracted many migrants from outside Spain, have made it possible to speak about increasingly glocalized urban impacts on rural hinterlands. Associated with this growing glocalization has been a significant redistribution of land. So, while years ago agricultural land was highly concentrated in a few farmers' hands, at present land is more equitably allocated, with many farmers sharing it. Hence, a lot of farmers now benefit from a purchasing power that is considerably higher than that enjoyed by only a few in the past. This has occurred hand-in-hand with substantial employment generation in the area, where the benefits of greenhouse agriculture have been based on the adoption of leading-edge technology. In this context, a lot of new jobs associated with agriculture have emerged, such as jobs related to the manufacture of plastic or iron for the building and maintenance of greenhouses, posts involved in the fabrication of agro-chemical fertilizers, engineering services, the production of machines and tools for agriculture, the development of irrigation systems, the making of cardboard and other containers for packaging, work in processing and marketing farm output, and, last but not least, activities associated with transport and storage. In addition, greenhouse production itself needs a lot of labour at all levels of productive chain. At the lowest levels people are employed on the hardest activities, which are those requiring more workers. Often, these activities are carried out by immigrants, who have been attracted to areas of labour shortage (Checa, 1995; Roquero, 1996), as compared to other parts of southern Spain, where the availability of indigenous workers has restricted openings for immigrants (e.g. Cruces Roldán and Martín Díaz, 1997). Part of the reason for immigration in the El Ejido area is that, along the coastline, tourism and associated developments in the hotel and restaurant sector, have increased. This has occurred at the same time as a form of 'agro-tourism' has spread, which consists of showing visitors greenhouses and what happens within them.

Employment changes in El Ejido have resulted in lower unemployment rates in its hinterland, where the general trend during the 1980s and 1990s was of a progressive rise in job numbers, at the same time that more diversification of occupational structures took place in the city-region. This happened in the context of unmet job demands, which arose from natural growth in the population and from the continual arrival of in-migrants. As a result, it has not always has been possible to record a shrinkage in jobless numbers, albeit, as with elsewhere in Spain, some of this is due to an unwillingness to accept farm work (e.g. Mansvelt Beck, 1988; Hoggart and Mendoza, 1999; Navarro, 1999). Certainly, despite demands for more jobs, there is socio-economic dynamism in the region, which has been brought about primarily by modern greenhouse agriculture.

Of course, employment generation has not occurred with the same intensity in all the city-region. Its strength has varied depending mainly on land availability and the fitness of local conditions for intensive agriculture. In this regard, El Ejido municipality has undergone the most important changes, followed by Adra and Berja. The least change has been experienced in Albuñol, which, to a large extent, is due to less appropriate local conditions.

The fact that modern agriculture has been the driving force in the growth of urban influences makes El Ejido a different case from most other European ones, where the spread of urban influence has involved a gradual falling off in agricultural activities. Yet the impact of this urban growth (and its spread effects in to the city hinterland) are similar to those outlined in the Granada city-region; as seen in a growing diversification in occupational structures, residential expansion and service improvements. To a certain extent, these improvements have been favoured by the fact that the urban development has demanded more facilities, in order to answer increasing demands resulting from population enlargement. As a consequence, there has been a huge rise in education, health, sport and cultural facilities, as well as in public transport, albeit these have taken place at a slower pace than demographic growth and subsequent urban sprawl. Not unsurprisingly this has given rise to social tension, conflict and imbalances, whose resolution is one of the main challenges for El Ejido's authorities. In addition, a series of services linked to modern agriculture, such as the transport of commodities, banks, recycling, etc., have developed. The number of companies devoted to these services has seen spectacular expansion in the last few years, with the tertiarization of the workforce also resulting from the spread of coastal and agrarian tourism, which has entailed a significant advance in the services related to it.

Similar to the areas analyzed around Granada, disparities exist between service levels enjoyed by residents and by visitors, for the distribution of services is not uniform across the El Ejido city-region. Hence, one finds disparities across social groups in the area; particularly between Spanish and immigrant communities. In addition, there are significant imbalances across the city-region, with areas less affected by agricultural modernization seeing services stagnating.

## Conclusion

This last section is devoted to stocktaking on changes that result from urban spread pressures in the case studies. The aim is to demonstrate how, independently of the causes of change, there are noteworthy common impacts for agriculture and spatial development, while such changes provoke, as a common outcome, socio-economic restructuring and revival in territories affected by them.

Firstly, concerning agriculture, in the hinterland of Granada a gradual drop in the importance of the sector has resulted from the spread of urban influences. In this regard, the enormous exodus of people from the core city to hinterland zones in the metropolitan region has brought great residential expansion, which, along with mounting demand for land for industrial uses, has given rise to a spatial disarticulation that has broken-up the traditional agrarian equilibrium that was characteristic of the fertile plain around Granada, named La Vega, where the traditional cultivation of tobacco is in decline as the profitability of farmland is progressively lowered, at the same time as financial yields from land devoted to urban uses rises ever more.

In particular, from the 1960s up to the present time, many isolated houses and residential areas have been developed along the road axis between Granada and

Monachil, most of which are located on former farmland. This tendency has been associated with a progressive fall in agricultural activity, such that, in places like Monachil, this activity is now of minor importance economically, being mainly focused on production for self-consumption, with low output and employment. A similar decline has been experienced in La Alpujarra, and with lower intensity in Güejar Sierra. Overall, many aged farmers are waiting for planners to designate their land as developable plots, so they have the opportunity to sell at higher prices.

Alongside this, tourism is giving rise to a redefinition of the socio-economic functions and cultural significance of rural spaces. In the cases studied here, this has been materialized in the reutilization of many traditional houses for residential and recreational purposes (this process is especially intense in La Alpujarra), at the same time as there is a continual increase in housing numbers and in the rehabilitation of cottages, even shepherd refuges. The aim is primarily to convert into accommodation or tourist facilities. The number of such buildings continues to grow in the zones analyzed here. In fact, whether in La Alpujarra or in other places where rural tourism is developing, this is more than a straightforward restoration of old dwellings, for what commonly occurs is a complete rebuilding of buildings. Such units often retain only their outdoor architectural image, since their interiors have been gutted, and the traditional dwelling house now has electricity, running warm water, heating, new furniture and other facilities that characterize modern homes. Tourists living in these houses now have the possibility of enjoying a supposedly traditional and rustic home, without renouncing the accustomed practices of their daily urban lives. They do not suffer the limitations and shortages that previous inhabitants of these houses bore regularly. What these rural tourists get is an imitation of the traditional home they inhabit.

Generally speaking, where rural tourism is a new activity that is replacing agriculture as the local economic base, there is modification to land-uses, and views on the 'function' of land. Far from being a requirement for farm production, that produced a scenery reflective of traditional ways of life, today rural space is viewed more and more as consumption space; that is, an arena in which consumers (that is, visitors or tourists) satisfy demands for leisure that are so characteristic in modern advanced urban societies (Hadjimichalis, 2003). As is well-known, such demands often value more isolated places highly, as these areas are associated with contact with nature, finding long-established and curious cultures, and so on (Camarero and Oliva, 2002). In this context, rural spaces are increasingly desired by urban dwellers, who are often disenchanted with their hectic and hyper-rationalized lives. These people are prone to 'run away' from their daily routines and yearn and/or look for territories that promise contact with nature and the 'exotic' (e.g. Urry, 1995). Of course, while the attractiveness to urban visitors of such territories is that they encourage new experiences and exciting sensations, what they really offer is mere simulations of the 'reality' that is expected by urban visitors. These simulations are idealizations regarding rural space that are grounded in the lived world of the visitors rather than of the longstanding local population.

Contact with the 'natural' and the 'authentic', which for many city people means rural tourism, can be materialized in the majority of cases only during occasional periods of leisure. This helps increase the idealization of their 'new rurality'. This is a sort of myth constructed by tired urban people who temporarily go to rural ambits to regain their strength. This myth has little to do with the 'new rurality' of longstanding rural inhabitants, many of whom have to adapt to new requirements of functional restructuring that embody far-reaching processes and decisions. These requirements bring with them the requirement that inhabitants shift their activities from essentially food producers to being responsible for the preservation of landscape and environment, as well as, in some cases, providing 'exotic' elements for the entertainment of urban visitors (e.g. Ilbery and Bowler, 1998; Halfacree, 1999).

Unlike Granada, in El Ejido, far from suffering decline, agriculture has developed enormously and is the key driving force in fast urban growth, the spread of urban activities outward from the city, and a more general dispersal of urban influence over the countryside. However, this is a particular type of agriculture, whose agro-industrial nature involves a land-use that is very different from traditional agriculture, which has virtually disappeared in this city-region. Yet, even in this zone, the consequences for farmland of urban spread effects are similar to those in Granada. Hence, at the same time as yields on greenhouse activity are peaking, escalating demand for land for construction outside El Ejido City is bringing a vast rise in land prices, even to the extent that new buildings now occupy spaces on which greenhouses previously stood. This is often accompanied by setting-up new greenhouses in the foothills of nearby mountains, whose relative distance from the urban core makes them less prone to the speculative interests of urban developers.

Similar to the effects of tourism in the city-region of Granada, the processes of intense change in El Ejido have entailed sizeable modifications in socio-economic functions and in the cultural significance of the countryside. The El Ejido city-region now has a scenery, socio-economic role and cultural significance that are very different from the past. In short, in the same way that territories around Granada are affected by growing urban pressures, what has happened in El Ejido city-region results from the exertion of urban pressure over previously rural areas. The key distinctiveness of this city-region is that urban pressure has been more intense, with huge quantities of building taking place, which has been materialized by the construction of more and more greenhouses and by widespread urban sprawl associated with fast growth of El Ejido City.

In general, change brought about by the extension of urban influence into the countryside has been more intense where, in addition to urban spill-over effects, forces of a global nature are prominent. This can be shown when we compare Güejar Sierra (an area seeing less globalized influences, where relatively modest changes can be attributed largely to spill-over effects from Granada City) with Monachil, La Alpujarra or El Ejido, where the international ski resort of Pradollano (including the worldwide diffusion of the Sierra Nevada image) and the expansion of modern agricultural practices to service an export market have been decisive in linking local development to what happens at a global scale.

Nevertheless, whether mutations basically result from urban spill-over or from connections with global change processes, a common outcome is socio-economic restructuring and the revival in territories affected by these forces In reality, globalization impacts are reflected more in the intensity of restructuring and in the pace of territorial revival, than in the consequences of all this change. These consequences are shown in land-use changes caused by the transition from traditional agriculture to an emphasis on rural leisure, or when the land is converted into space for agro-industry, as happens in El Ejido. Independent of the causes and intensities of changes that accompany urban spill-over effects, for the areas investigated, a common impact from these changes is a progressive increase in the attractiveness of the areas for outsiders, whether they aim to work, live or visit. This attractiveness has a twofold effect. On the one hand, it is a key factor provoking greater urban pressure and socio-economic dynamism. On the other hand, this dynamism is one of the main reasons why these territories are confronted by new social problems. Such social problems are often an effect of inequalities suffered by territories impacted on by the aforesaid dynamism. These inequalities are shown when, in the areas investigated in Granada, we compare different social situations, such as those of traditional farmers that remain and people in new service or manufacturing occupations, just as it is in the living conditions of newcomers and 'traditional residents'. Regarding El Ejido, the unevenness is specially seen in the superior standards of living of autochthonous people and of in-migrants, especially if these are illegal immigrants.

Concerning origins, imbalances can be a consequence of the new circumstances brought about by widening urban influences or can result from a broadening of previous disparities, which are triggered by urban spread influences. Whatever their origin, a common outcome of urban spread processes is a trend toward the aggravation of inequalities, which provides evidence that the ESDP goal of achieving balanced development is not being fulfilled as these processes progress. However, deepening inequalities do not happen with the same intensity in all areas affected by urban spread effects. This is not completely explainable if one only interprets urban effects as an expression of the diffusion of influences from established urban cores. In this regard, as demonstrated in this chapter, the exacerbation of inequalities is especially intensified in rural territories where urban effects occur hand in hand with growing glocalization and mounting socio-economic dynamism, such as has taken place in Monachil, La Alpujarra and, above all, in El Ejido.

## References

Camarero, L.A. and Oliva, J. (2002) Urban-rural turnaround and the changing shape of Utopia, Paper presented to the *XIIIth World Congress of the International Economic History*, Buenos Aires, Argentina, Session 35: *Explaining Counter-Urbanization: Historical Approaches to Urban-Rural Migration*, Convenor: Jeremy Burchardt, 1-15

Champion, A.G. (1989, ed.) *Counterurbanization*, Edward Arnold, London

Checa, F. (1995) Oportunidades socioeconómicas en el proceso migratorio de los inmigrantes africanos en Almería, *Agricultura y Sociedad*, 77, 41-82

Commission of the European Communities (1999) *ESDP – European Spatial Development Perspective: Towards Balanced and Sustainable Development of the Territory of the European Union*, Office for Official Publications of the European Communities, Luxembourg

Coombes, M.G. (2000) Defining locality boundaries with synthetic data, *Environment and Planning*, A32, 1499-1518

Crecente, R., Miranda, D., Cancela, J. and, Marey, M. (2001) Potencialidad de la ordenación parcelaria para la multifuncionalidad del espacio rural, Paper presented to the *IV Spanish Portuguese Colloquium for Rural Studies* on *La Multifuncionalidad de los Espacios Rurales de la Península Ibérica*, Santiago de Compostela, 7-8 June, 1-21

Cruces Roldán, C. and Martín Díaz, E. (1997) Intensificación agraria y transformaciones socioculturales en Andalucía occidental: análisis comparado de la costa noreste de Cádiz y el condado litoral de Huelva, *Sociología del Trabajo*, 30, 43-69

Fernández, A. (2004) Turismo y ordenación del territorio, Online journal *Quaderns de Política Econòmica*, Department of Applied Economics, University of Valencia, 7, 36-47

García Sanz, B. (1994) Nuevas claves para entender la recuperación de la sociedad rural, in *Papeles de Economía Española*, Madrid, 60-61, 204-218

García Sanz, B. (1997) *La Sociedad Rural ante el Siglo XXI*, Ministerio de Agricultura, Pesca y Alimentación, Madrid

Hadjimichalis, C. (2003) Imagining rurality in the New Europe and dilemmas for spatial policy, *European Planning Studies*, 11, 103-113

Halfacree, K.H. (1999) A new space or spatial effacement: alternative futures for the post-productivist countryside, in N. Walford, J. Everitt and D. Napton (eds.) *Reshaping the Countryside: Perceptions and Processes of Rural Change*, CAB International, Wallingford, 67-76

Hoggart, K. and Mendoza, C. (1999) African immigrant workers in Spanish agriculture, *Sociologia Ruralis*, 39, 538-562

Hoggart, K. and Paniagua, A. (2001) The restructuring of rural Spain?, *Journal of Rural Studies*, 17, 63-80

Huylenbroeck, G. van and Durand, G. (2003, eds.) *Multifunctional Agriculture: A New Paradign for European Agriculture*, Ashgate, Aldershot

Ilbery, B.W. and Bowler, I.R. (1998) From agricultural productivism to post-productivism, in B.W. Ilbery (ed.) *The Geography of Rural Change*, Longman, Harlow, 57-84

Mansvelt Beck, J. (1988) *The Rise of the Subsidised Periphery in Spain*, Nederlandse Geografische Studies 69, Utrecht

Marchena Gomez, M. (1984) *La Distribución de la Población en Andalucía 1960-1981*, Diputacion Provincial de Sevilla y Universidad de Sevilla, Sevilla

Navarro, C.J. (1999) Women and social mobility in rural Spain, *Sociologia Ruralis*, 39, 222-235

Paniagua, A. (2002) Counterurbanization and new social class in rural Spain: the environmental and rural dimension revisited, *Scottish Geographical Journal*, 118, 1-18

Pellicer Corellano, F. (1998) El ciclo del agua y la reconversión del paisaje periurbano en las ciudades de la Red C-6, in F. Javier Monclús (ed.) *La Ciudad Dispersa*, Centro de Cultura Contemporánea de Barcelona, Barcelona, Chapter 5

Precedo Ledo, A. (1988) *La Red Urbana*, Síntesis, Madrid

Precedo Ledo, A. (1996) *Ciudad y Desarrollo Urbano*, Síntesis, Madrid

Romero, J. and Albertos, J.M. (1993) Retorno al sur, desconcentración metropolitana y nuevos flujos migratorios en España, *Revista Española de Investigaciones Sociológica, (REIS)*, 63, 123-144

Roquero, E. (1996) Asalariados africanos trabajando bajo plástico, *Sociología del Trabajo*, 28, 3-23

Salmon, K.G. (1995) *The Modern Spanish Economy: Transformation and Integration into Europe*, second edition, Pinter, London

Sancho-Martí, J. (1989) *El Espacio Periurbano de Zaragoza*, two volumes, Ayuntamiento de Zaragoza Cuadernos de Zaragoza 59, Zaragoza

Sjoberg, G. (1964) The rural-urban dimension in preindustrial, transitional and industrial societies, in R.E.L. Faris (ed.) *Handbook of Modern Sociology*, Rand McNally, Chicago, 127-159

Tout, D. (1990) The horticulture industry of Almería Province, Spain, *Geographical Journal*, 156, 304-312

Urry, J. (1995) A middle-class countryside, in T. Butler and M. Savage (eds.) *Social Change and the Middle Classes*, UCL Press, London, 205-219

Zárate, A. (1984) *El Mosaico Urbano: Organización Interna y Vida en las Ciudades*, Cincel, Madrid

Chapter 6

# Tensions, Strains and Patterns of Concentration in England's City-Regions

Steven Henderson

**Introduction**

Underpinning the European Spatial Development Perspective's (ESDP) vision of competitive city-regions is more effective integration of cities with their hinterlands (Commission of the European Communities, 1999). Rather than conceiving of hinterland areas in a somewhat unidimensional fashion, as a land bank, the ESDP calls for more extensive rural-urban integration and co-operation. Thus, with agricultural markets and the rural economy experiencing fluctuating fortunes, the ESDP states that: 'The future prospects of surrounding rural areas are also based on competitive towns and cities' (Commission of the European Communities, 1999, p.22). Not only can stronger rural-urban links help create new markets, but opportunities may exist for off-farm employment and agricultural diversification (Ilbery, 1988). Simultaneously, an open and accessible countryside may encourage businesses to relocate into the wider city-region. In terms of managing urban expansion on to rural land, the ESDP advances effective land-use planning, including urban containment, as a means of reducing the need to travel and the environmental impact of sprawling growth. The ESDP also stresses a need to enhance the quality of life of hinterland residents. In summary, for hinterland areas, the document calls for: 'Integrating the countryside surrounding large cities in spatial development strategies for urban regions, aiming at more efficient land-use planning, paying special attention to the quality of life in the urban surroundings' (Commission of the European Communities, 1999, p.25). Linked closely to notions of sustainable development, the ESDP's hope for the city's hinterland is one in which economic, environment and social objectives are balanced in a fashion that delivers wide-ranging benefits.

Reflecting on the English context, this chapter recognizes that reducing urban sprawl has been a long-term policy objective. This noted, the introduction of sustainable development principles into the land-use planning system appears to be reinforcing the objective of urban concentration. At the same time, there is evidence that rural-urban linkages are strengthening, as indicated by the intensity of rural to urban commuting. Whilst this provides partial confirmation that England is conforming to the broad principles contained within the ESDP's vision, concern is raised for the quality of life of hinterland residents. In particular two anxieties are presented. The first relates to the management of housing growth, which cannot be contained within existing built-up areas. Here tensions are identified in the way that large-scale urban housing growth has been directed towards designated rural expansion settlements. Whilst this represents an example of stronger co-operation between rural and urban local authorities, as encouraged by the ESDP, a key dilemma for such settlements is whether they are viewed as independent, or must now look toward nearby centres for services. To an extent this mirrors tension in the public policy literature between the needs of a city-region and the quality of life of rural communities that must accommodate urban functions that have spilled out from metropolitan centres (Popper, 1985). A second concern relates to the quality of life of residents in smaller rural settlements, where strong rural-urban links between larger rural towns and urban centres undermines the strategic function of such towns in servicing surrounding villages. Here the ESDP offers ambiguous policy advice. On the one hand, it supports stronger rural-urban interaction, indicating that small- and medium-sized towns can play a key role in '... easing accessing to the bigger labour markets' (Commission of the European Communities, 1999, p.22). On the other, it highlights the role of larger rural centres as focal points for regional development.

By exploring the existing literature and presenting new empirical analysis, this chapter examines land-use change and related strains within hinterlands zones in England. The first section highlights the tensions that exist within rural land-use planning, not only between local residents and urban newcomers, but also between policies seeking to concentrate development and locally identified needs (Elson, 1986). The emergence in UK public policy of sustainable development principles during the past decade, including the promotion of 'urban renaissance', restrictions on greenfield development and a stress on the need to reduce in green house gas emissions (UK Department of the Environment, Transport and the Regions, 2000), have provided renewed emphasis on the geographical concentrating of development. Having identified the emergence of such tendencies in the first section of the chapter, the second section draws on almost 100 in-depth interviews with planners, housing officers, community organizations, councillors and business groups to investigate the implications of this emphasis. To strengthen conclusions the chapter draws on government publications and numerical evidence, including population statistics and the results of a household sample survey. The primary focus for the project as a whole was the city-regions of Bedford, Cambridge and Norwich within the East of England region, although the spotlight in this chapter is on the Norwich city-region.[1] Differentiating between expansion settlements and those hinterland villages where the prospects of growth are limited, this section highlights the tensions that are emerging. The final section acknowledges that, whilst there is evidence of compliance with the ESDP in the English context (e.g.

---

1  The project from which this research was drawn is 'Urban pressure on rural areas (NEWRUR)', which was funded by the European Commission, under its Quality of Life and Management of Living Resources Programme (contract QLK5-CT-2000-00094).

Faludi, 2003), there remains a need for stronger policy coherence to meet the needs of hinterland residents.

## Growth Management Within the City's Hinterland

Research linking urban areas to land-use change in their wider hinterlands has been relatively limited in the UK, as well as internationally (Ford, 1999). In part this stems from the success of the UK planning system in creating a strong and often abrupt division between urban areas and the countryside (Elson, 1986). Thus, the urban literature has focused on built-up areas, and the rural literature has a strong orientation towards smaller settlements and the countryside. Although the rural landscape around UK urban areas may appear less fragmented than wider international experience (Cloke, 1989), there are nevertheless strong connections between rural and urban areas. One sign of this is the considerable attention that has been given to the experiences of farmers within the city's hinterland, including tensions associated with their urban proximity (Munton, 1974; Munton *et al.*, 1987; Ilbery, 1988; Catherine Bickmore Associates, 2003). Offering a somewhat different perspective, in this section of the chapter the focus is more on house-building within city-regions. Here the intention is twofold: first, to recognize that policies supporting the concentration of housing in designated areas have been less than successful because of concern for local needs; and, second, that the emergence of sustainable development principles has renewed inherent emphases on the geographical concentration of new development.

Managing the expansion of urban areas in the UK has been a problem that planners and government officials have grappled with for well over a century. During the 1940s a modern planning system emerged emphasizing the creation of land-use plans to guide growth to pre-designated sites and protect the countryside for agricultural production. To help with urban containment, many English cities witnessed the designation of greenbelts during the 1950s, following the lead established around London in 1938 (Munton, 1983). In combination such policies helped control the outward expansion of metropolitan areas, thus maintaining a sharp divide between rural and urban areas and preventing them coalescing through urban sprawl. Pressure on England's urban fringe zones has remained strong, with the literature highlighting the intensity of development pressures that can emerge just beyond the city's suburbs, including strategies to encourage or limit the conversion of farmland (Munton, 1983; McNamara, 1984; Pacione, 1990). The tourniquet-like effect of green belts has encouraged criticism where economic activity is felt unreasonably constrained (Tewdwr-Jones, 1997) or where housing development leapfrogs over protected zones (Monk and Whitehead, 1996), resulting in long-distance commuting and road congestion (While *et al.*, 2004). Further questions have been raised about the environmental quality of land protected by green belts (Elson, 1986) and whether green wedges have greater planning merit (Tewdwr-Jones, 1997). Yet compared to the international experience, where development pressure has carved its way through greenbelt designations (e.g. Freestone, 1992), in the UK, despite small-scale losses, greenbelts have largely been protected, and have strong public and political support.

Where house building on the urban fringe has been constrained, how to direct future housing growth has been an important concern for government officials (Gilg, 1991). Rather than condoning the dispersal of housing pressure, the tendency has been to designate selected areas for growth. Thus, for larger metropolitan areas, a series of new towns and town expansion schemes were designated to cope with urban overcrowding and the need to redevelop dilapidated inner-city areas between the 1950s and 1970s (Wannop, 1999; Hall, 2002; Pacione, 2004). Although attempts were made to ensure that such settlements did not function simply as dormitory suburbs, the results were mixed in terms of their ability to generate a stable employment base (Lawless and Brown, 1986). The capacity of new towns to ease prevailing housing pressures has also been criticized (in much the same way as in the Netherlands; Ostendorf, 2001). In the period 1961 to 1989, London's 11 new towns (including Milton Keynes, Peterborough and Northampton) absorbed only a fifth of the growth experienced within the Greater South East Region outside of Greater London (Hall, 1989; Wannop, 1999). Thus, despite attempts to deliver a concentrated form of deconcentration, new settlements did not prevent development in and around towns and villages. Even greenbelt policies, which prohibited settlement expansion, permitted infilling and rounding-off, if local approval could be obtained (Elson, 1986). In rural areas more generally, housing growth received support because of a belief that additional housing could strengthen the existing community, and thus help protect against the loss of services (Elson, 1986).

Where development has spilled outward from cities into the countryside, policy-makers have historically attempted to concentrate development around larger rural settlements, where a more extensive range of services can be accessed (Shucksmith *et al.*, 1993). Yet past experience in concentrating activity has not always proved successful. Even in terms of new towns, Wannop (1999, p.229) recognized the '... occasional shortages of schools, shops and some social facilities'. Similar concerns have emerged where key rural settlements have been designated to act as centres around which housing, services and employment concentrate. In addition to providing alternatives, so rural residents did not have to migrate towards larger metropolitan and regional centres, key settlements were designated to absorb housing pressure in heavily populated rural areas. By concentrating development in key settlements, it was held that services could be provided more efficiently and economically, and that a desirable alternative could be presented to rural households experiencing difficulty in accessing services and employment (Cloke, 1979). Despite some success in bringing living and work spaces closer together, and improving access to services, the key settlement policy met various implementation problems (Blacksell and Gilg, 1981; Philips and Williams, 1984). For Cloke (1979, p.217): 'The reliance on restrictive rather than positive planning in rural areas has meant that vital stages in the development of key settlements and in the spread of opportunities to their hinterlands are absent from present planning'. Ultimately, the absence of well thought through alternatives were manifest in an 'official' tolerance of housing expansion in rural settlements than policy guidance anticipated (Cloke, 1979; UK Department of the Environment, 1988).

Rather than simply a response to housing supply, the outward movement of urban pressure has been encouraged by a strong cultural preference for living in the countryside (Macnaughten and Urry, 1998). This preference has increased demand for newly constructed properties as well as for existing dwellings. Hardly surprisingly, the migration of urban residents into commuter villages, and the social transformation experienced within such settlements, has been a longstanding research issue (e.g. Pahl, 1965; Connell, 1974). Research exploring the interaction of urban areas and their surrounding rural commuter belt has, however, tended to fray over time. Gradually, whilst rural research remained grounded within particular localities, 'the urban' has been dealt with at a more generalized level. As the influence of London over the South East of England intensified, the capital's influence overlapped with the hinterlands of smaller urban areas, thus adding to the complex nature of rural-urban interactions. Once it came to be recognized that few parts of rural Britain were immune from large-scale urban originating in-migration, whether for commuting, second home ownership or for permanent residence (Cross, 1990), the emphasis within the literature changed from exploring the effects of urban proximity to a stress on examining how rural areas were being 'restructured' (Marsden *et al.*, 1993).

Representing something of a disjuncture from the pattern of dispersal identified above, the literature reports that once urban residents have relocated into a rural village they commonly seek to restrict future house building. Even where they benefit from new housing themselves they may wish to raise the 'drawbridge' on future housing and employment developments, in order to protect local architectural heritage, to preserve environmental quality and to safeguard land values (Murdoch and Marsden, 1994). Yet in terms of community attitudes towards future housing growth, protests from newcomers must be balanced against regional variation in urban pressure (Marsden, 1995), that the demand for housing may be higher in attractive settlements, that housing sub-markets exist, such as stronger demand for period properties, and that newcomers hold different attitudes to longstanding residents (Cloke and Thrift, 1990; Cloke *et al.*, 1994; Mormont, 1987). Equally important are protest lodged by longstanding residents toward large-scale housing developments, for reasons including infrastructural limitations and architectural quality (Hedges, 1999). Where urban households have moved into the countryside, the existence of rural-urban income differentials, and the ability of in-migrants to out-bid existing residents, has meant that urban pressure on rural areas is often associated with a housing affordability problem for local households (Bramley and Smart, 1995; Hoggart, 2003). This transition has been hastened by national housing policy, including the 1980 Housing Act, which provided existing tenants with the 'right to buy' public sector housing at subsidized prices, with subsequent re-sales offering a further entry point for middle class rural home ownership (Chaney and Sherwood, 2000). For many the force with which urban pressures fall on rural settlements raises questions about the effectiveness of mechanisms through which social housing is provided (Hoggart, 1997, 2003). This, however, masks the wider question of whether social housing should be provided in rural villages, particularly given reports that rural services, including public transport, are notoriously poor. It also gives inadequate attention to the preferences of and the location-based opportunities extended toward low-income and more immobile rural households.

Thus far it has been noted that attempts to accommodate housing needs through concentration in key settlements have proved less than successful, and that pressure for housing growth to meet local needs remains strong in hinterland settlements that are integrated into urban housing markets. With the emergence of sustainable development principles during the 1990s, concentrated development received renewed support. Arising out of various international declarations, sustainable development principles seek to satisfy the needs of current generations without compromising the livelihoods of future inhabitants (World Commission on Environment and Development, 1987). The underpinning priority is the need for a more environmentally benign human existence, with safeguards for existing resources and biodiversity. Whilst its importance as a strategic priority continues to be debated (Owens and Cowell, 2002), particularly in view of technological change, sustainable development has been adopted by successive UK governments as the key underpinning of the land-use planning system. As interpreted in the UK, sustainability policies promote the re-use of brownfield sites over greenfield land (UK Parliamentary Office of Science and Technology, 1998), and the creation of a settlement pattern than minimizes the emission of greenhouse gases (UK Department of the Environment, Transport and the Regions, 2000). For the latter, reducing carbon dioxide emissions from private cars is a key priority, leading to the promotion of more balanced and concentrated forms of development, where housing, employment and services are provided in close proximity. This policy stance requires stronger integration of land-use and transport planning so new housing has strong public transport links to employment and service nodes (Banister, 1994). Similar priorities have been expressed at the European level in the European Spatial Development Perspective (ESDP), which calls for a balanced and sustainable development of the territory of the European Union. Incorporated into the ESDP are five key dimensions for the sustainable development of towns and cities; namely, controlling their physical expansion, reducing social exclusion, protecting eco-systems, environmentally friendly transport options and the conservation of natural and cultural heritage (Commission of the European Communities, 1999, p.22). As in the UK, the ESDP supports the notion of the compact city, the re-use of derelict industrial land and development patterns that minimize the need to travel. To provide further reflection on the UK Government's latest attempt to concentrate development, its impact on hinterland settlements and whether contradictions and tensions are emerging, the remainder of this chapter focuses on the Norwich city-region.

**Development Patterns Within the Norwich City-Region**

In reviewing the recent experiences of hinterland settlements in the Norwich city-region, this section is divided into three parts. The first offers an historical context for land-use by describing patterns of economic growth and housing development. The next two sections focus on Norwich's hinterland settlements, investigating villages that have seen significant housing growth in recent decades, followed by those whose prospects of housing expansion are restricted.

By way of introduction, the Norwich built-up area[2] lies within the county of Norfolk, located some 180km north east of London. With a population of 193 000 in 2001, Norwich is an important regional centre, which lies at the heart of a vast rural area (Corfield, 1994). The county is characterized by a network of radial roads extending outwards from Norwich to a ring of market towns, which are surrounded by their own smaller hinterlands. Norfolk's strong agricultural tradition is reflected in a high density of rural villages, and by the expansion of commercial and industrial activity within Norwich and surrounding market towns to service the farming economy and process farm produce. Largely unaffected by the Industrial Revolution, the economy has experienced modest growth with fortunes rising and falling depending on the strength of agricultural markets. Because of its distance from London, population growth has been modest historically. However, between 1991 and 2001, the resident population increased by 5.0 per cent. This is higher than the 4.3 per cent growth experienced across England and Wales, but is comparatively less than England's fastest growing settlements, such as the Cambridge build-up area, which experienced a growth rate of 7.2 percent.

It would be wrong to infer from this description that Norwich has not experienced recent periods of economic restructuring (Perkins *et al.*, 2000). The economic recession that gripped the national economy in the early 1990s was felt in Norwich. By 1996, the city was in the economic doldrums, with suggestions that the absence of key urban infrastructure, '... poor transport links, job losses at Nestlé, Norwich Union and the HMSO [Her Majesty's Stationery Office] have left a city short of confidence and with major industrial sites derelict' (*Eastern Daily Press*, 1996). Out of this decline, Norwich's economy has undergone revival. Building on operations in the insurance sector, it has witnessed strong growth in financial and business services. The expansion of its retail and leisure sector has in turn strengthened Norwich's position as the regional centre of an expansive hinterland. Evidence of Norwich's (and to a lesser extent Norfolk's) transformation from a manufacturing-based economy to a stronger services presence can be seen in Table 6.1. Stemming from this restructuring, the registered unemployed fell from a Norfolk peak in February 1993 of 35 665 to 9 185 in July 2003 (Norfolk County Council, 2003, p.4).

Despite its distance from South East England, Norfolk witnessed pressure spilling outwards from London and the Home Counties during the second half of the 20th century. By the late 1960s, the post-war pattern of young adults and elderly people moving from London to surrounding counties (Warnes, 1991) had extended into Norfolk. As a result the rate of house building increased from 1 000 to

---

2  For clarity, a distinction is made between Norwich and Norwich City. This is because the Norwich built-up area is fragmented between three local government authorities: Norwich City Council, Broadland District Council to the north and South Norfolk District Council to the south.

approximately 8 000 per annum. The development pressure that emerged was felt strongly in Norfolk's rural villages, with many transformed from rural hamlets to suburban villages (Beckett and Madgett, 1993). Thus, Chinery (2000, p.20) indicates that, whilst 1 per cent of new houses were built in villages with less than 500 inhabitants in the 1950s, by the 1970s this figure had reached 20 per cent. To manage such pressure, Norfolk County Council adopted its first rural settlement policy in the 1960s. This plan saw development boundaries drawn around approximately 300 villages, outside which expansion was not supposed to occur. However, rather than being seen as delimiting the zone beyond which development would not occur, these boundaries were interpreted as indicating where housing would be approved, so Norfolk's villages were hit with a further wave of 'suburbanization'. Initially housing growth was strongest in villages close to Norwich and along major trunk roads, but by the 1970s virtually all villages within 80 km of Norwich experienced housing growth (Norfolk County Council, 1980).

**Table 6.1    Percentage of Norfolk Employment by Economic Sector, 1991-2001**

|  | 1991[a] | | 2001[b] | | |
| --- | --- | --- | --- | --- | --- |
|  | Norwich City | Norfolk | Norwich City | Norfolk | East of England |
| Agriculture, forestry, fishing | 0.44 | 4.94 | 0.81 | 3.87 | 1.92 |
| Energy and water | 1.21 | 1.50 | 0.40 | 0.52 | 0.63 |
| Mining and quarrying | 0.95 | 1.21 | 0.15 | 0.52 | 0.21 |
| Manufacturing | 19.13 | 17.66 | 13.08 | 15.19 | 14.47 |
| Construction | 7.15 | 8.25 | 6.88 | 7.91 | 7.62 |
| Distribution and retail, | 21.99 | 23.06 | 23.44 | 23.57 | 21.45 |
| Transport | 5.62 | 4.96 | 5.26 | 5.43 | 7.41 |
| Banking and finance | 12.63 | 10.40 | 19.19 | 14.43 | 19.12 |
| Government and other | 30.08 | 26.94 | 29.97 | 28.56 | 27.17 |

Source:    Office of Population Censuses and Surveys (1994) and UK Office of National Statistics (2003)
Notes:     [a] Percentage of employees and self-employed residents
           [b] Percentage of people aged 16-74 employed

Since the late 1970s, a key principle in emerging planning policies has been the need for a better balance between housing and employment. Thus the 1980 Norfolk Structure Plan states that: 'The scale and location of new housing development will, in the 1980s, be related increasingly to the rate and location of job growth in the county. The greatest potential for job growth is in and around Norwich' (Norfolk County Council, 1980, p.1). Outside Norwich key settlements were identified around which attempts were made to balance employment, housing and services. Beyond these settlements, within the '… vast majority of small villages there will be no significant change in their physical appearance, with any new development only on a small scale' (Norfolk County Council, 1980, p.3). For Norwich and its immediate hinterland, a Norwich Policy Area was identified to draw Norwich over-spill back toward the city, both by designating urban fringe sites for expansion and by selecting proximate hinterland settlements for growth. The partial effectiveness of the policy is evident in Table 6.2. Significant growth in Broadland reflects a policy of directing Norwich's population expansion to the northern fringe of the city. This policy rests on a longstanding determination to protect the Yare River Valley, which lies just south of the City. In contrast to suburban growth, the inner-city of Norwich City declined. Yet Breckland, in the midst of rural Norfolk, stands out for its rapid growth, which contradicts the policy of concentrating housing in proximity to Norwich.

**Table 6.2   Population Growth Rates for Norfolk Local Authorities, 1981-2001**

|  | Mid-year estimate 1981 | Mid-year estimate 1991 | 1981-1991 change | % change 1981-1991 | 2001 | 1991-2001 change | % change 1991-2001 |
|---|---|---|---|---|---|---|---|
| Breckland | 96 700 | 108 300 | 11 600 | 20.5 | 121 400 | 13 100 | 35.1 |
| Broadland | 98 100 | 107 200 | 9 100 | 16.1 | 118 500 | 11 300 | 30.3 |
| Great Yarmouth | 81 400 | 88 900 | 7 500 | 13.3 | 90 800 | 1 900 | 5.1 |
| KL & W [a] | 122 000 | 131 700 | 9 700 | 17.2 | 135 300 | 3 600 | 9.7 |
| North Norfolk | 83 300 | 92 000 | 8 700 | 15.4 | 98 400 | 6 400 | 17.2 |
| Norwich | 126 100 | 127 200 | 1 100 | 1.9 | 121 600 | -5 600 | -15.0 |
| South Norfolk | 95 200 | 104 100 | 8 900 | 15.8 | 110 700 | 6 600 | 17.7 |
| Norfolk | 702 900 | 759 400 | 56 500 |  | 796 700 | 37 300 |  |

Source:   Norfolk County Council (2002)
Note:     [a] King's Lynn and West Norfolk

Focusing specifically on South Norfolk, the themes above are illustrated in Table 6.3. Of note is generally rapid growth during 1961-1971. This is in contrast to population decline in the rural policy area and in 11 of the 31 District parishes in the Norwich Policy Area in the previous decade. Population growth from 1961 to 1971 was most intense in the northern Norwich Policy Area.[3] During the following inter-censual period, the rate of growth declined, and the balance of population change shifted further from the City toward the southern rural area – with a high percentage outside the three main rural service settlements. It was not until

---

3   The Norwich Policy Area consists of Norwich and (as interviewees called it) 'a necklace' of surrounding settlements, which are designated for growth. This Area is smaller than what can be considered Norwich's hinterland or the Norwich city-region.

1991-2001 that a higher rate of District growth again occurred in the Norwich Policy Area. This 1991-2001 pattern adheres to the policy of concentrating development closer to Norwich. In addition to growth in designated settlements within close proximity to Norwich, the rapid expansion of the market town of Wymondham, located immediately to the south west of Norwich, is evident. Growth within the rural policy area nevertheless remained strong for 1991-2001. Whilst 40 per cent of population change was concentrated in the main three rural settlements, housing development in rural villages continued apace. Across the county, between 1993 and 2002, 60 per cent of Norfolk parishes had up to one new dwelling completed per year, appropriately a third had between one and 10 dwellings a year, and less than 10 per cent of parishes had over 10 dwellings added yearly (Norfolk County Council, 2003, p.3). With the number of completions in rural parishes exceeding the figure in market towns and larger rural centres combined, Norfolk County Council concluded that whilst development was concentrating, there was scope for this to be intensified.[4]

**Table 6.3  Population Growth Within South Norfolk Settlements, 1951-2001**

|  | 1951 population | Population change | | | |
|---|---|---|---|---|---|
|  |  | 1951-1961 | 1961-1971 | 1971-1981 | 1981-1991 | 1991-2001 |
| Norwich Policy Area | 24 064 | 3 159 | 10 481 | 6 170 | 4 329 | 4 660 |
| Norwich suburbs | 5 863 | 2 312 | 2 318 | 328 | 989 | 88 |
| Hinterland growth settlements | 4 865 | 713 | 4 504 | 4 134 | 2 229 | 2 183 |
| Market town [b] | 5 665 | 239 | 2 609 | 1 300 | 903 | 1 823 |
| % of change in above three |  |  | 90.0 | 93.4 | 95.2 | 87.9 |
| Rural Policy Area | 39 222 | -1 555 | 5 324 | 6 329 | 4 384 | 4 143 |
| % change in largest three centres [c] |  |  | 24.3 | 29.1 | 44.3 | 42.1 |
| South Norfolk | 63 286 | 1 604 | 15 805 | 12 499 | 8 713 | 8 803 |

Source: Norfolk County Council (2005)
Notes: [a] Settlements experiencing the highest 1961-2001 growth rate were Easton, Hethersett, Long Stratton, Mulbarton, Newton Flotman, Poringland and Tasburgh. On average, Easton, Newton Flotman and Tasburgh expanded by 232 per cent from 369 to 1 226. In Hethersett, Long Stratton, Mulbarton and Poringland the average population increased from 1 118 to 3 807 in this period.
[b] Wymondham
[c] Diss, Harleston and Loddon

---

4   Similar patterns can be identified in other local authorities. Norfolk County Council figures indicate that mid-1991 to mid-2001 Breckland house completions totalled 2 344 in larger urban areas (Thetford and Dereham), 1 697 in rural service centres (Attleborough, Swaffham and Watton) and 1 855 in rural parishes. In Broadland, outside the Norwich Policy Area, which saw 6 150 new dwellings, house completions totalled 448 in the designated rural centre of Aylsham and 1 488 in rural parishes.

During the early 1990s, the search for new housing land within the Norwich Policy Area fell increasingly to the south of the City in South Norfolk District. With northern suburbs facing increasing traffic congestion, the need to rebalance the spatial configuration of the City became self-evident (Norfolk County Council, 1994). For rural settlements to the south, the construction of a southern by-pass (completed in 1992) improved links to the City and provided a stronger argument for expansion to the south. With South Norfolk District Council facing pressure to accommodate additional housing, a key debate emerged over whether to designate a new village (to be called Mangreen), which would accommodate 3 000 people or to spread growth. South Norfolk District Council voted for the latter (*Eastern Daily Press*, 1990, 1990a). To help accommodate this, Norfolk County Council Structure Plan (1993), extended the Norwich Policy Area to the south beyond the first ring of settlements, to encompass Long Stratton, which lies approximately 16 km from the centre of Norwich. Stemming from this change, Long Stratton's population increased from 2 896 to 3 701 during the 1991-2001 inter-census period. In this period, growth was also experienced in the hinterland settlements of Hethersett, Newton Flotman and Poringland. However, on balance, stemming in part from a time lag between planning policy and housing construction, the highest rates of population growth continued to be in Broadland (Table 6.4). Analysis of Norfolk County Council's house completion figures for mid-1991 to mid-2001 reveals a similar pattern. During this time the rate of house completions across the three local government areas were registered as Broadland 8 086, Norwich 2 586 and South Norfolk 4 136.

**Table 6.4  Population Growth Within the Norwich Policy Area, 1981-2001**

| | | 1981 | 1991 | % change 1981-1991 | 2001 | % change 1991-2001 |
|---|---|---|---|---|---|---|
| Norwich City | | 122 890 | 122 661 | -0.2 | 121 550 | -0.9 |
| Broadland | | 98 417 | 104 846 | 6.5 | 118 513 | 13.0 |
| Norwich Policy Area | Built-up | 50 666 | 52 503 | 3.6 | 61 945 | 17.9 |
| | Remainder | 15 725 | 18 279 | 16.2 | 19 780 | 8.2 |
| Rural Policy Area | | 32 026 | 34 064 | 6.4 | 36 788 | 8.0 |
| South Norfolk | | 93 194 | 101 907 | 9.3 | 110 710 | 8.6 |
| Norwich Policy Area | Built-up [a] | 10 821 | 11 810 | 9.1 | 11 898 | 0.7 |
| | Remainder | 33 053 | 36 393 | 10.1 | 40 965 | 12.6 |
| Rural Policy Area | | 49 320 | 53 704 | 8.9 | 57 847 | 7.7 |

Source:   Norfolk County Council (2005)
Note:     [a] Costessey and Cringleford

From the mid-1990s, development agendas altered significantly nationally and within the Norwich city-region. In some ways replicating Norfolk's longstanding policy of directing housing toward employment growth, national planning guidance stressed achieving more sustainable development. Planning Policy Guidance (PPG) 13 on transport, released by the UK Department of Transport in 1994, highlighted the need to '… move towards a better balance between employment and population, both within existing urban areas and in rural communities in order to enable people to live near their work' (paragraph 3.5). In the case of Norfolk the need to re-examine past planning policies was given further weight by the continuing deconcentration of people and by heightened commuting to Norwich. The latter is evident in Table 6.5. Norwich City is the only local government area within Norfolk with a positive net-commuter flow. The highest rates of out-commuting are from the surrounding districts of Broadland and South Norfolk, with significant out-commuting from the rural heart of Norfolk (e.g. Breckland). Of particular concern for those seeking a strategic local balance between housing and employment, job opportunities in larger market towns have not kept pace with dwelling construction, so resulting in out-commuting from market towns toward Norwich. Thus, as one county administrator concluded:

> … from free-standing market towns, they are now largely residential areas, which have continued to grow rapidly simply because housing is cheaper. They then have quite long journeys to work into Norwich. Some sites have attracted some small-scale light industry, but really not enough to offset the decline in the surrounding agricultural employment, coupled with some of the traditional industries in the towns.

**Table 6.5   Residents, Workplaces and Daytime Populations Aged 16-74 in 2001, by Norfolk District**

|  | Population | | | Commuter volumes | | |
|---|---|---|---|---|---|---|
|  | Resident | Workplace | Daytime | Inward | Outward | Net |
| Breckland | 86 783 | 45 299 | 76 470 | 9 537 | 19 850 | -10 313 |
| Broadland | 86 322 | 39 230 | 67 588 | 14 789 | 33 523 | -18 734 |
| Great Yarmouth | 64 808 | 36 192 | 63 416 | 7 465 | 8 857 | -1 392 |
| KL & WN [a] | 97 398 | 56 356 | 93 557 | 8 414 | 12 255 | -3 841 |
| North Norfolk | 70 438 | 37 452 | 66 479 | 7 135 | 11 094 | -3 959 |
| Norwich | 89 832 | 92 558 | 128 784 | 53 752 | 14 800 | 38 952 |
| South Norfolk | 79 883 | 39 898 | 67 155 | 13 885 | 26 613 | -12 728 |

Source:   Norfolk County Council (2004a, p.3)
Note:     [a] King's Lynn & West Norfolk

This is particularly noticeable for Dereham, with Norfolk County Council (2002a, p.31) stating that this market town '… is still largely dependent on Norwich for jobs, leading to long distance commuting'. Using selected Norfolk settlements, a sense of how widespread this pattern is can be derived from Table 6.6, with rural settlements such as Loddon and Aylsham having noticeably high rates of out-commuting to Norwich. Of further importance is the high usage of private vehicles for travelling to Norwich, both within the Norwich Policy Area and from rural service towns. Further evidence of job concentration in Norwich is provided by a job density indicator. Calculated by the Office for National Statistics (2003a), this is a ratio of filled jobs in an area to the working aged residential population. With a job density ratio of 1.31 in 2001, Norwich City has the highest ratio of jobs per person of any local authority within the UK outside the inner-city London boroughs of City of London, Westminster, Camden, Kensington and Islington. This highlights that Norwich has stronger linkages to its surrounding rural districts than many other regional centres, such as Cambridge (1.24), Exeter (1.13), Ipswich (1.07), Oxford (1.06) or York (0.99). A concentration of jobs in Norwich is further reflected in traffic volumes converging on the city centre. In its Local Transport Plan, Norfolk County Council (2004, p.42) acknowledged that: 'Norwich has chronic traffic congestion, with traffic moving more slowly than in any other UK urban area apart from two. This is mainly because Norwich functions as a much bigger city, but only has the infrastructure of a small city'. This is exacerbated by travel for leisure. Norwich is one of the top 10 retail centres in the UK (Norfolk County Council 2002a), yet the smallness of its resident population compared to

**Table 6.6 Patterns of Commuting to Norwich from Selected Settlements, 2001**

| | Employed and self-employed people aged 16-74 | % of workers aged 16-74 working in Norwich | % travelling to Norwich by | | |
|---|---|---|---|---|---|
| | | | Car | Car passenger | Bus |
| Norwich Policy Area | | | | | |
| Hethersett | 2 729 | 45.44 | 76.91 | 7.98 | 11.11 |
| Mulbarton | 2 218 | 47.70 | 79.60 | 7.42 | 8.34 |
| Poringland | 2 015 | 50.62 | 77.38 | 8.45 | 8.31 |
| Stratton | 2 342 | 32.15 | 77.32 | 7.73 | 12.37 |
| Tasburgh | 1 555 | 40.95 | 79.70 | 8.18 | 8.48 |
| Wymondham | 6 143 | 28.58 | 75.39 | 7.73 | 11.00 |
| Rural service towns | | | | | |
| Aylsham | 3 045 | 26.31 | 79.74 | 7.80 | 8.43 |
| Dereham | 7 067 | 14.83 | 77.13 | 8.63 | 10.87 |
| Loddon | 1 211 | 31.71 | 79.38 | 11.00 | 7.56 |
| North Walsham | 5 156 | 15.74 | 74.67 | 7.11 | 12.00 |

Source: Calculated using the web interface of the Census Interaction Data Service using ward level figures (http://cids.census.ac.uk)

lower ranked retail centres indicates that it is highly dependent on its rural catchment. For a former senior County Council Planning and Transport officer:

> I mean Norwich is different. I go back to this ratio between the urban core and the dependent rural hinterland which is more extreme than almost any other city in Western Europe. And they always say it is biggest. Well, they argue about Inverness and Aberdeen, which have large dependent rural populations relative to the town. But they are smaller towns than Norwich. You will find very few with virtually a million population looking to an urban core of 200 000.

A key reason for this is that recent growth in the Norwich built-up area is strongly oriented toward the city centre. One exception is growth on Norwich's southwest fringe, near the University of East Anglia, which has had notable recent developments, including the construction of the new Norfolk and Norwich Hospital. Yet, despite its recent good fortunes, Norwich's economic growth is not generally reflected in the outward migration of businesses. The exception is the emergence of warehouse or distribution operations in out-of-town business parks. More generally, commercial estate agents note that only in recent years has it become viable to convert former farm buildings in close proximity to Norwich into business activities. For businesses, the central city retains its attractiveness. Central to this is the recent regeneration of Norwich's core, including an expansion of retail and leisure outlets, and more recently the redevelopment of inner-city brownfield sites for housing. For similar reasons, the likelihood of businesses wishing to relocate to market towns was seen as very unlikely. Thus, according to one senior City Council official:

> When this major business park at Thorpe St Andrew [east of Norwich] was first proposed [see Valler, 1996], we were actually very concerned that it would suck the life out of the city centre and we would get a lot or relocation. ... And in fact it didn't happen at all. There were virtually no relocations. So I think what that shows is the inherent attractiveness of the city centre.

The impact of sustainable development priorities in Norfolk has thus been to redirect housing growth towards Norwich. Because of time lags in housing construction, change is most evident in policy priorities. In Norfolk's 1993 Structure Plan, which covered 1988 to 2006, 32 per cent of new housing was to be in the Norwich Policy Area (Norfolk County Council, 1993). In the 1999 Structure Plan, which covers 1993 to 2011, the figure increased to 35 per cent (Norfolk County Council, 1999). The introduction of Planning Policy Guidance 3 (PPG3) on housing provided further weight for redirecting growth toward Norwich. PPG3 states that '... local planning authorities should follow a search sequence, starting with the re-use of previously-developed land and building within urban areas identified by the urban housing capacity study, then urban extensions, and finally new development around nodes in good public transport corridors' (UK Department of the Environment, Transport and the Regions, 2000, paragraph 30). Based on this logic, recent Structure Plan review documents propose further increasing this figure, possibly to 45 per cent (Norfolk County Council, 2002a). The concentration of future housing closer to Norwich is also evident in estimates of population growth. To reinforce a point made earlier about re-balancing

Norwich's historical northward growth, South Norfolk is predicted to experience the highest growth from the 1990s up to 2011, absorbing 26.5 per cent of Norfolk's population expansion (Norfolk County Council, 2000). In accommodating future housing it appears that PPG3 is helping direct a higher proportion of dwellings toward the edge of Norwich's built-up area, as opposed to nearby hinterland settlements. This contrasts with the picture presented in Table 6.4. Thus, in planning for Norwich's housing requirements in the period 2001 to 2006, South Norfolk's Local Plan designates an additional 66 hectares for development in the Norwich Policy Area. Representing a change from the pattern identified in Table 6.3, 42 hectares is located on the edge of Norwich (Costessey 25.15, Cringleford 13.00, and Trowse with Newton 4.20), with a further 23.60 hectares in settlements with good connections to the City (Poringland 12.00, Wymondham 6.41, Easton 2.70, Long Stratton 2.25) (South Norfolk District Council, 2003).

Underpinning PPG3 is the principle that wherever possible the housing needs of an urban centre must be concentrated within the existing built-up area. Only then should planning authorities look toward urban fringe land release, and only finally towards nearby settlements with good public transport links. In this urban-centred logic additional housing in smaller rural villages is viewed negatively as it encourages commuting with private vehicles. Yet there remains ambiguity within current government advice, because policy statements raise the potential for small-scale housing developments. Thus, PPG3 states that '… only a limited amount of housing can be expected to be accommodated in expanded villages' (UK Department of Environment, Transport and the Regions, 2000, paragraph 70), with priority given to villages falling on transport corridors, local service centres where there is an absence of nearby alternatives, and villages where '… additional houses are needed to meet local needs, such as affordable housing, which will help secure a mixed and balanced community'.

Norfolk, because of intense rural-urban links between Norwich and its rural hinterland, it appears that priority is given to reducing the intensity of commuting rather than a broader assessment of village needs. For example, within the 2003 South Norfolk Local Plan, the Government Inspector overturned the Council's proposal to allow housing growth in 28 settlements, reducing this number to 12. Principally, these 12 had good communications with Norwich or were larger settlements. Thus, one village of more than 3 000 inhabitants, two of 2 000-2 999, six of 1 000-1 999 and seven of under 1 000 inhabitants were excluded when the 28 became 12. In making this ruling, the Inspector concluded that villages in South Norfolk were strongly linked to Norwich, so growth would strengthen their dormitory function. As Table 6.5 highlighted, approximately 50 per cent of South Norfolk's working population commutes out of the local authority. The trend toward concentration is captured in the summary provided by a housing officer in South Norfolk District Council:

> ... the last local plan had land allocated for housing, pretty well in every village of any size at all, and it was scattered all around the District, 40-50 villages. This time it is big sites in the larger settlements and it is a very distinct policy change. It was imposed on us really by the local plan inspector. We had got a lot more sites scattered around in the small villages because the councillors wanted to keep the possibility of people staying in their own village. The local plan inspector took the sequential approach and threw out lots of our other sites. So we have now got 700 in Poringland, 1 500 in Costessey, 300 in Cringleford.

In the discussion above a trend toward concentration within the Norwich city-region has been highlighted. Within this broad trend, distinctions do nonetheless need to be drawn between two settlement categories: those that have seen substantial housing growth in recent decades; and small villages that are likely to see only limited new housing. In the next two sections the tensions and strains that are prevalent in both of these hinterland settlement types are reviewed.

**Tensions Within Norwich's Expansion Settlements**

This section focuses on those hinterland settlements around Norwich that have experienced significant population growth in recent decades. As housing pressure has spilled outward from Norwich, land in these settlements has been designated for additional housing. The effect has been that such communities have become increasingly dependent on Norwich for employment and services. At one level, it can be argued that this provides evidence of interaction between rural and urban areas, as promoted by the ESDP. Yet, stemming from this integration, a particular anxiety exists over whether such settlements are mere suburbs of a physically expanding Norwich or represent independent communities requiring additional services. As seen by local governmental and non-governmental agencies, in reality community needs within these settlements are overlooked by government agencies. They are seen to fall through a gap between key public policy initiatives for rural and urban areas. In urban areas, the dominant focus is toward the regeneration of the inner-city, whilst significant attention and resources in rural areas are directed toward strengthening market towns.

A common concern in interviews with local councillors was that new housing estates are attached to hinterland settlements with insufficient attention to future service requirements or pressure on existing services. The absence of services was seen to reinforce the dormitory nature of such settlements, with newcomers using villages and small towns as places to sleep while their waking (work and leisure) lives are oriented toward Norwich. Of particular concern was service shortages, including libraries, educational facilities, village halls, sports buildings and playing fields. This anxiety is depicted in the following two comments:

> It's dumping, that is what they have done. They have dumped them [new houses] on us. And it wouldn't be so bad if they gave something with it. But they don't. (Poringland councillor)

> Poringland... Hethersett, I always regard it as very similar, large villages effectively just being nothing more than dormitory suburbs for Norwich. Hethersett is now the

same it has now got 5 000 people. Its bank has just closed. It's got about two shops and a school and nothing else. (Social housing officer)

This tension is not unique to Norwich's hinterland settlements, for concern that housing estates are constructed with insufficient attention to the services required to create integrated communities is a problem recognized nationally. As Kearns and Turok (2003, p.37) comment: 'Many housing estates built during the 1980s and early 1990s on the edge of cities amounted to little more than monotonous suburban sprawl. They were "soulless" communities lacking facilities, vitality and quality design'. Confirming the significance of this observation, the Government recently stated that: 'We cannot divorce this housing issue from the wider challenge of creating more successful, more sustainable communities. We need to learn from the lessons of the past, when we simply built dormitory towns without sufficient jobs, public services or other facilities' (UK Office of the Deputy Prime Minister, 2005, p.3). As PPG3 Housing states: 'Any substantial new development, whether a town extension, village expansion or new settlement should not consist exclusively of housing but must be planned as a community with a mix of land uses, including adequate shops, employment and services' (UK Department of Environment, Transport and the Regions, 2000, paragraph 66). Whilst similar in nature to suburban development, hinterland settlements do differ as they can easily (perhaps even conveniently) fall into the gap between 'rural' and 'urban'.

Interviewees gave various reasons for the poor servicing of Norwich's expanding hinterland settlements. First, national and local planning policies are strongly oriented toward the protection and regeneration of urban centres. By increasing the attractiveness of inner-city living, it is thought that pressure for greenfield land on the urban fringe can be decreased. Significant attention is therefore given to the re-use of brownfield land, including former industrial sites, to create an urban renaissance (UK Department of the Environment, Transport and the Regions, 2000b). To encourage this transformation, in an age of intense inter-city competition (Harvey, 1989), emphasis is given to iconic buildings and urban spectacles that attract business interest, property investors and tourism. Norwich's adoption of this strategy is evinced in the transformation of derelict land adjacent to the train station into a large mixed-use development (David Simmonds Consultancy, 2004) and in investment to enhance visitor appeal, such as the new City library (The Forum). Arguably, as acknowledged by the two South Norfolk councillors, directing public funds toward central Norwich has limited the dispersal of services to hinterland settlements:

> And libraries, we have just built this huge library in Norwich, but really the argument is, we are near enough to use that. But [...] it is very costly to do so. And it is not an argument that people around here would understand. I mean they want to form a community in its own right here. And I think that is healthy ... It is lack of funds really, or lack of will, or lack of recognition of the need, but more who is going to pay for it all. (South Norfolk councillor)

> We keep being told what do you want a library for, you are only six to seven miles [10-11 km] from the Forum [Norwich City Library]. What about the old girl [retired

woman] that can't get there on the bus, and we have got a lot of them. (South Norfolk councillor)

To provide further evidence of the concentration of investment in Norwich, a review of the distribution of National Lottery funding since 1995 reveals that 59.42 per cent of Norfolk's allocation was directed to agencies and organizations based in Norwich City. At least 60 per cent of this figure was spent on major Norwich-based projects including the Forum, an East Anglian Sports Park, a new inner-city swimming complex, Norwich Castle Museum, a new record/archive office, and the restoration of historical buildings and a local cathedral.[5] The focus on inner-city Norwich is reinforced by a determination on the part of local authorities to protect the city centre as the heart of the wider region, and the focal point toward which roads and public transport are oriented. For the same reason, planning officials acknowledge that, compared to other city-regions, Norwich had managed to limit car-oriented out-of-town retail developments, except for supermarkets selling food products. The same logic has been applied to private leisure-based land-uses, such as cinemas or recreation facilities, with planning officials prioritizing the protection of existing facilities in central locations. For this reason there are reservations about permitting the widespread distribution of such services, at least until demand is more certain. As acknowledged by a County Council planner:

> Three or four years ago there seemed to be lots and lots of sports and leisure clubs, and again we were concerned about the unrestricted growth of out-of-town sports and leisure clubs might be on things like our major swimming pools. [...] But a lot of these things, when the trend first emerges, when the demand is unclear, clearly if you are only going to get one or two or three examples, then if they don't go in the city centre, then it is not a good idea, but if you are going to get many, many of them, and the demand is that great that you are going to get lots of them, then some kind of dispersal is a good thing.

A second identified reason for the inadequate servicing of Norwich's hinterland is that many government agencies focus on improving market towns. According to the Rural White Paper, the government must provide '... support to develop the potential of Market Towns for their economic role (including leisure and tourism) and as service centres' (UK Department of the Environment, Transport and the Regions, 2000c, p.12). Underpinning this strategy is the Countryside Agency's Market Towns Initiative, which receives funding from regional development agencies, although, as with many government initiatives, funding is constrained. Illustrative of this, the East of England Development Agency scheme only supports approximately four market towns in each of the five counties under its jurisdiction. Looking beyond this initiative, only 16 Norfolk settlements outside the Norwich Policy Area are designated as rural growth or rural service centres in the current Norfolk Structure Plan. Notably this excludes Norfolk's five largest centres: Dereham, Great Yarmouth, King's Lynn, Norwich and Thetford (Norfolk County

---

5   Calculated using national lottery grant figures accessed from the Department of Culture, Media and Sport's website, http://www.lottery.culture.gov.uk/.

Council, 1999). But funding is not restricted simply by place, but also for initiatives. This is potentially significant, for the Market Towns Initiative involves public consultation to identify community needs and for the delivery of strategic priorities. But because funding, whether from regional government agencies or from local authorities, was not always forthcoming, this led to anxiety that some centres are doing better than others. As a councillor for a rural settlement with just under 3 000 people stated:

> We would love to get a swimming pool. They have just built one at Wymondham, so I don't think we are getting one. I think they have got an awful lot in Long Stratton as a direct result of the council offices being there ... You can get a bit cynical about it can't you.

With public policy focusing on market towns, attention is directed away from hinterland settlements despite comparatively similar sizes.[6] A review of National Lottery funding for organizations based in South Norfolk District for the period since 1995, focusing on grants allocated towards local needs, provides some support for this anxiety. Compared to a rate of funding of £33.9 per person for the market town of Wymondham, the figures for the hinterland settlements of Hethersett, Mulbarton and Poringland were £4.5, £18.4 and £268.4, with the rural service towns of Diss, Harleston and Loddon receiving £51.9, £52.4 and £27.6 respectively. In this example, Poringland stands out as an anomaly because of a £1.1million grant from the New Opportunities Fund for a new sporting complex. Allocated in 2004, after interviews were completed, to some extent this prize contradicts some local views, as captured by this South Norfolk councillor:

> ... they have also been thwarted by the East of England sport people, Sport East of England, I think, because I think there were saying, well, Poringland is only three miles [5 km] from Norwich, and you've got all the facilities there, why do we need to build something so close.

Yet, to ignore such concerns does risk downplaying the breadth of concerns relating to service under-funding and the time frame over which lobbying is required to obtain additional services. This point was echoed by one of Poringland's councillors:

> ... we have a village hall ... but it is totally inadequate for an area of this size. And there is a little bit of money from the Parish taxes to pay for a replacement, but that has been an ongoing battle for 12-14 years, and still no sign of it coming about. There is no help from District Council for it at this point. It is a long drawn-out thing, but, you know, nobody is rushing forward to really provide these things.

Similar funding concerns were raised in the case of Wymondham. As one representative from a local development partnership acknowledged, recent funding from the Countryside Agency for a development officer was surprising:

---

6  Taking South Norfolk as an example, 2001 ward populations for Hethersett, Mulbarton and Poringland were 5 441, 3 261 and 2 827, respectively, compared to 6 917, 4 058 and 2 578 for three designated rural service towns of Diss, Harleston and Loddon.

> We have worked very hard over the last 10 years to bring Wymondham back onto anybodies map of the world ... it has traditionally been a working town, hasn't had a large articulate middle class, it has not been good at speaking up for itself ... the notion of commuting from Wymondham has been well established. So we easily get lumped in with Greater Norwich to our detriment. For example, thinking of the Countryside Agency and the market towns initiative, Wymondham in one sense isn't a market town.

A third reason for poor services within expanding hinterland settlements is the seemingly low level of financial contributions collected from house builders to offset growth pressures on local services. Under the authority of Section 106 of the Town and Country Planning Act, 1990 (as amended by Section 12 of the Planning and Compensation Act, 1991), local authorities are permitted to enter into legally binding agreements with developers for the purpose of restricting or regulating development or a proposed land-use change in order to achieve relevant planning objectives, provided that they relate to the impact of, or the pressures created by, the particular development in question, rather than fulfil wider community needs. Prior to this, contributions from developers could be sought through Section 52 of the Town and Country Planning Act, 1971 (see Bunnell, 1995 for a review). Whilst restricted to the off-site impact of developments, there was evidence in Norwich's hinterland suburbs of developers offering to fund community services on the back of housing proposals. In the case of Wymondham the offer of sporting facilities to address existing shortages led to tension between environmental interests, who were keen to protect local amenity, and local sporting interests. Labelled as little more than a bribe by South Norfolk District Council, the application was refused (*Evening News*, 1988; Poynton, 1988).

In line with national trends, local authorities have in recent years become more demanding in negotiating Section 106 contributions from developers (Campbell *et al.*, 2000). Associated with increasing local authority pressure for new developments to fund service improvements, authorities have a growing list of desired contributions. To avoid strain on existing services and facilities, contributions may be requested for fire hydrants, schools, libraries, open space, pedestrian paths, roads and other community facilities. Of particular importance, given house price inflation across Norfolk since the mid-1990s (see Table 6.7), have been contributions toward affordable housing. But in the development industry, a common concern is that, if contributions continued to increase, the pace of house building will slow as profits are constrained (House Builders' Federation, 1998). With claims that the financial viability of house building is limited, community and government priorities tend to be traded-off, seemingly in a rather haphazard process of negotiation. Campbell and associates (2000) make a similar point in terms of planners having to balance the competing needs of local authority departments. In the words of a planning officer from South Norfolk:

> In Costessey we are getting 6 per cent [affordable housing]. We could have said, no, we are going to go for broke, we want 25 per cent, and they would have said, right, you don't get the schools, you only get half a road, you get no open space.

In terms of prioritizing developer contributions, tensions are evident between the preferences of local residents and the priorities of local authorities. In hinterland settlements, settlement function is at the core of this tension. One particular friction relates to whether contributions should improve public transport to Norwich, thus reinforcing patterns of dependency on the regional centre, or go toward local facilities and community projects. As suggested by the following local councillor, there is a sense that money has been directed toward enhancing rural-urban connections rather than local social priorities:

> There certainly wasn't any negotiation before some of the agreements were made, ... £170 000 has gone to the bus company to alter the service from a 20 minute service to a 15 minute service for the next two years. So at the end of the two years, they can either say it is viable, in which case they are making money anyway, or it is not viable and they are going to stop it. But to change it from 20 minutes to 15 minutes is stupid in the first place.

**Table 6.7   Average Change in Property Prices for Selected Norfolk Local Authorities, 1999-2004**

| Local authority district | Average property price (£) 1999 | Average property price (£) 2004 | Increase (£) 1999-2004 | Percentage increase 1999-2004 |
|---|---|---|---|---|
| Norwich City | 60 935 | 141 735 | 80 800 | 132.6 |
| Broadland | 74 009 | 170 701 | 96 692 | 130.6 |
| South Norfolk | 79 595 | 184 824 | 105 229 | 132.2 |
| Breckland | 67 943 | 156 380 | 88 437 | 130.2 |
| Norfolk | 69 618 | 158 698 | 89 080 | 128.0 |

Source:   Land Registry Property Prices, http:www.landregistry.gov.uk

One response to poor service provision within Norwich's expanding hinterland settlements is a trend toward larger developments, from which planners can demand more comprehensive developer contributions. For instance, as the scale of housing estates increases, a case can be made for developers to provide a sports field rather than a contribution to a smaller open space, or potentially a whole school as opposed to a future site. This is highlighted within Norfolk County Council's recent Structure Plan review. Looking across a 25-year span, the document recognizes the potential to plan for major mixed-use developments. In addition to ensuring that developments contain necessary infrastructure and community facilities, emphasis is given to creating '... an identifiable community rather than suburban sprawl' (Norfolk County Council, 2002a, p.38). In suggesting three possible locations for a new major development, the north east of Norwich fringe was favoured rather than significant expansion in Wymondham, as the latter is seen to be more likely to encourage car usage, despite its proximity to employment growth on Norwich's south west fringe. The option of a new village was criticized for being '... poorly related to existing jobs and services and unlikely to attract businesses so as to be truly 'mixed-use' (Norfolk County

Council, 2002a, p.39). Again this highlights the difficulty economic development officials have in encouraging the decentralization of employment from Norwich.

## Strains Within Norwich's Smaller Commuter Villages

This section focuses on smaller villages in Norwich's commuter belt, which in recent decades have seen incremental expansion, but whose future growth is likely to be constrained by the adoption of sustainable development principles. A key concern in the literature about such settlements is that, as newcomers arrive, housing becomes unaffordable for local people (e.g. DTZ Pieda Consulting, 1998). Although interviewees supported this view, the main argument presented in this section is that, whilst the notion of socially mixed villages is political desirable, the underlying reality is that affordable housing for low income and immobile residents is poorly located relative to available services. There needs to be a strengthening in market town and larger rural centre roles in offering services, employment and housing for surrounding villages. Without comprehensive policies, current trends will result in stronger links with Norwich, so downplaying the potential role of larger rural centres as foci for surrounding villages.

Early sections of this chapter have highlighted that, whilst development is increasingly concentrated within the Norwich city-region, the construction of housing in villages continues, with the total number of village completions exceeding those in market towns and rural service centres combined. The possibility that sustainable development principles constrain future housing growth receives a mixed reaction. In view of the fact that South Norfolk has experienced rampant house price inflation during the past decade (see Table 6.7), in the mind of at least one local councillor: 'I would like to see a cul-de-sac in all of the villages with say 10 new houses'.[7] Often it is anticipated that such development would provide opportunities for local house buyers, help preserve local services and create options for households who live in inappropriate housing to remain locally. Yet for various reasons this viewpoint is subject to intense criticism. Three main reasons are presented. First, where housing has been developed on small plots in rural villages this is commonly large executive dwellings, which are unaffordable for local residents. Second, planning officers question the advisability of housing expansion in smaller villages, doubting that there is a correlation between the availability of rural services and housing growth. From an historical perspective it is noted that rural services have declined during periods when housing growth occurred in villages (Shaw, 1980). More anecdotally, the impact of newcomers on services draws a mixed response. Whilst some commentators report that commuters are more likely to shop within Norwich or out-of-town supermarkets (see Findlay *et al.*, 2001), others note that newcomers are more vocal in protesting about the quality of village services. Importantly, the latter is viewed positively

---

7 This sentiment has been expressed in national circles. For example, the last Countryside Agency chairperson, Ewen Cameron, called for six or eight new homes in every village, rather than large greenfield housing estates (*The Guardian*, 2004).

when it involves community support for childcare facilities but negatively when it results in pressure for pedestrian paths and additional lighting.

Third, planning officers are highly critical of additional housing in smaller settlements, because few jobs are available locally, so residents need to commute to access employment. With only 33 per cent of rural Norfolk households having an hourly or better bus service within 13 minutes walk (Norfolk County Council, 2003, p.7), this encourages travel by private vehicles, so contradicting sustainable development principles. The flipside is that, whilst UK government documents, such as the Rural White Paper (UK Department of the Environment, Transport and the Regions, 2000c), are supportive of rural employment growth, including farm diversification, sustainable development principles question the advisability of local opportunities that are poorly serviced by public transport. Thus, in the case of an application to replace and extend an existing bakery in a rural parish in South Norfolk, and so expand the number of employees from 14 to as many as 50, in recommending refusal the relevant planning officer stated that: 'There are clearly vacant sites within the identified built up areas of Diss and Harleston, within 13km and 12km of this site which could satisfy the site requirements of the new bakery without compromising normal policy constraints' (South Norfolk District Council, 2002, p.5.2). Although this ruling was overturned by the Council, it highlights a tendency toward centralization in planning system interpretations. The impact of this on rural investment requires further investigation. Already it appears that sustainability principles encourage tension between economic development officers and their planning counterparts, as highlighted in the following comment by a Norfolk County Council development officer:

> So, if Government policy is not to allow any kind of development to happen there at all, then it becomes increasingly difficult to justify diversification schemes that a farmer might put forward in one of those villages. So it does seem as if the Government just wants to channel everyone into as few a number of viable centres as they possible can.

What the above also highlights is potential tension between councillors and planning officers, with the former showing concern for the social needs of villages, while the latter place greater weight on contemporary environmental agendas. This is not to suggest that there is widespread local support for housing in small villages. In a rural social survey of 471 households in Breckland and South Norfolk,[8] in response to an open-ended question asking about concerns for the future of their village, 38.7 per cent were perturbed by the idea that they would see more new housing. Anxiety was also expressed by councillors when additional housing means the loss of village services. For instance, a review of planning control data from South Norfolk District between 1992 and 2001 revealed that 17

---

8   Twenty-three villages were examined in population groups 'under 1 000', '1 000-2 000' and 'more than 2 000' (the largest village crept above 3 000 by 2001). Settlements also varied in their urban linkages, as indicated by the percentage of paid-workers commuting to the Norwich built-up area (the nearest centre of 50 000 inhabitants plus).

public houses (including five within rural service centres and market towns) had been lost for the creation of 28 additional dwellings.

A separate issue from the release of land for housing is the potential to construct affordable housing. Norfolk County Council (2002a, p.32) recently stated that: 'Building significant numbers of new houses in villages is environmentally unsustainable, but there is a case for affordable housing where there is an identifiable need'. With private sector housing schemes in rural villages usually too small to invoke a developer contribution toward affordable units (Norfolk County Council, 2002a), the main avenue through which dwellings can meet local needs is the rural exceptions policy. This policy enables a local authority to '... grant planning permission for small sites, within and adjoining existing villages... which the local plan would not otherwise release for housing, in order to provide affordable housing to meet local needs in perpetuity' (UK Department of the Environment, Transport and the Regions, 2000, Annex B, paragraph 2).

Whilst policies are in place to give people of all social classes the option of living in the countryside, local authority housing officers, rural housing enablers and housing associations, indicate that the exceptions policy is not conducive to the construction of many houses (see Hoggart and Henderson, 2005). As identified in the literature, difficulties relate to participants believing the development process takes an inordinate time, costs being higher than expected, fears about who eventual tenants will be and NIMBY-style opposition to proposals (Yarwood, 2002; Gallent and Bell, 2000). Two particular perspectives are evident from our research. First, of particular note in Norfolk, the limited success of rural exception policies has less to do with local opposition than the unwillingness of official agencies to promote the exceptions option (see Hoggart and Henderson, 2005, for greater detail). With policies now concentrating employment and services in a few centres, planning officers question the logic of placing low-income households in villages. Of further concern was that such houses would be on the edge of a settlement, not necessarily close to village services. A more significant blow to the objective of a socially mixed countryside is whether there is 'real' demand for rural exception sites. Local need is notoriously difficult to estimate, with officials not uncommonly criticizing local housing surveys for overestimating 'real' demand for homes:

> Woodton particularly is very strong on the fact that there are no affordable housing in their village, as far as they are concerned, and their children are having to go and live in Norwich, and work somewhere else, and can't stay in their village any longer. How real it is, is a moot point. About 12 years ago they did the survey, in the early days of affordable housing when people were looking, and came up with a requirement that there was six people in Woodton, who would stay in Woodton if there were affordable housing. Six separate families, or couples or whatever they were. So on the basis of that we got the Council to provide some land, free of charge, and Hastoe Housing Association, they built six houses. ... When it actually came to being built, we couldn't find any of those six. (South Norfolk councillor).

Our own household survey in rural Breckland and South Norfolk casts doubt on the level of 'real' demand for rural housing amongst village residents, with only

8.5 per cent of respondents (to an opened-ended question) citing an affordable housing shortfall as a cause of concern for the future of their village.

Planners see stronger links between rural villages and larger service centres as a desirable solution to the problems facing rural villages, yet the strength of linkages appear limited. As Norfolk County Council (2003, p.8) has recently commented: 'Market towns as a whole appear to be performing well, and fulfilling their role as service centres for their rural hinterlands', yet at best this is based on a narrow assessment of their retail economies, which reveal a low rate of shop vacancy. A broader perspective suggests that links between service centres and surrounding villages are more tenuous, which is strengthened by the influence of Norwich as the regional centre. In terms of the household survey, non-employment related travel to Norwich appeared strong in settlements of all sizes, with at least one household head travelling to the regional capital once a week for non-work purposes (Table 6.8). As for employment, for settlements with less than 1 000 people, 41.8 per cent of household heads work in Norwich, 45.1 per cent do so in settlements of 1 000-2 000 inhabitants and 39.4 per cent do so in settlements with more than 2 000. A Norwich focus is further evinced in 43.2 per cent of households responding to an open-ended question about what should be improved within the wider Norwich region by naming transport. Yet attention to the quality of transport links with Norwich detracts from the need for stronger linkages between villages and nearby service centres. As reported earlier, because of the centralization of employment opportunities in Norwich, rural households look towards the regional centre for employment rather than to nearby rural service centres. In 2002, outside of Norfolk's four largest centres [King's Lynn, Great Yarmouth, Norwich and Thetford], only six settlements had a town bus service, namely Dereham, Diss, Downham Market, Fakenham, North Walsham and Swaffham (Norfolk County Council, 2002b). Whilst in some cases bus services are provided from market towns to nearby villages, the coverage of these services is limited. As a district council local transport officer acknowledged:

> There is a group of people in Wacton and Great Moulton and Aslacton, where they have been pushing for a bus service for quite a long time. And the County [Council] have put on a bus service between that area and Wymondham, and it was lucky to have one or two people on it a day, which is painfully low.

Given this predicament it is little surprise that councillors look negatively on local people being forced to move to nearby larger towns. In the absence of public transport services, there is little difference between having to move two, 10 or 20 km away from existing friends and family to find housing.

**Table 6.8    Percentage of Non-Work Trips to Norwich During the Past Month by Village Population Size**

|  | Settlement size | | |
| --- | --- | --- | --- |
|  | Less than 1 000 inhabitants | 1 000-2 000 inhabitants | More than 2000 inhabitants |
| No trips | 8.7 | 9.7 | 8.9 |
| One trip every two weeks | 24.0 | 27.2 | 25.3 |
| At least one trip per week | 25.0 | 27.2 | 28.1 |
| One to two trips per week | 17.3 | 18.4 | 21.2 |
| Two to three trips per week | 14.0 | 13.0 | 8.9 |
| More than three trips | 11.3 | 8.7 | 7.5 |

Source:    Household survey

The servicing role of market towns also appears limited in terms of an ability to satisfy housing needs for surrounding villages. For one, the national Housing Corporation, which directs finance toward social housing construction, defines the cut-off for its special 'rural' programme as settlements with less than 3 000 people; a figure well below that of many market towns (Hoggart, 2003). As more than one social housing officer acknowledged, this seems to conflict with the national government's rhetoric of promoting market towns:

> ...they have moved towards looking at schemes in market towns that meet rural housing needs. They say that, but in practice they want the schemes in the rural areas, settlements under 3 000, and they are very tight on that... once a settlement goes above 3 000.

One avenue through which affordable housing can be delivered is as a planning contribution. Nationally, the increasing reliance on Section 106 agreements to provide social housing can be seen from the conclusion that almost one half of affordable units constructed in 2003 had some level of developer contribution (Monk *et al.*, 2005). By concentrating new rural owner-occupied homes in larger rural settlements, Section 106 contributions can help fund affordable housing. At the same time rural householders might benefit from their proximity to services.

In terms of households purchasing new owner-occupied dwellings in service centres, real estate agents observe that this market is oriented toward first time home buyers and Norwich over-spill. It is reported that this is not commonly due to households being forced to buy in larger centres rather than villages, as many prefer this option. Thus, as one Dereham estate agent commented:

> We have almost seen over the years this change. If I go back 20 years, people with young families used to want to live in the village, for the quality of village life. Now people with young families want to be in the town, because they are fed up with acting as a taxi service to football, ballet, all sorts of things that kids do these days. And there is nothing like that happening in the villages, it is nearly all based around the towns.

Interviewees in the housing field also noted unmet demand in villages for homes in larger centres. In particular, they see an absence of suitable properties for people who are physically less mobile. This is particularly evident in bungalows for older residents. Thus, as one rural estate agent concluded: 'We see quite a few older people, retired people, in their 70s who would like to move into the town, but we can't find a property to sell to them'. The limited development of such properties is seen to reflect national pressure for higher densities in new housing estates,[9] the impact on developer profits of more spacious bungalow developments and the ability of developers to evaluate and respond to market needs independently. An additional concern is that it is not always easy to obtain social housing in nearby centres because of the way the allocation systems work. This reflects the priority given to homeless households or those with serious social problems, compared to rural households experiencing accessibility and mobility problems. Pressure for social housing is also strong in many market towns, as implied by the following local authority social housing officer:

> So, in terms of housing need, Dereham is the most popular area. Attleborough we have pressures there to, mainly because I think Attleborough is the highest priced area in our District, with Dereham probably coming a close second.

In view of this demand to live in larger rural centres, a key question for the future is whether people will request exceptions housing in villages, not because they want to live there, but because housing opportunities elsewhere are constrained.

That there is a noticeable absence of governance structures oriented toward satisfying community needs within rural towns and their hinterlands has already been noted in the literature (Edwards *et al.*, 2003). Whilst surrounding villages are noted in market town health checks, poor levels of community engagement raise questions over whether rural needs are being considered (Sawyer, 2002, 2002a). Equally important is slight community pressure for a stronger rural employment base.[10] To take the case of Wymondham, anxiety emerged in the mid-1960s that the town was being forced to accommodate housing over-spill from Norwich without attention to employment growth, which raised town concerns about becoming a dormitory suburb (*Norwich Mercury*, 1965, 1971). Today, despite continuing population growth (Table 6.3), employment concerns are not at the forefront of community development strategies. As a representative from the Wymondham Development Partnership indicated:

---

9 Compared to more than 50 per cent of new housing that is currently being built with a density of less than 20 dwellings per hectare, PPG3 requires local authorities to obtain between 30 and 50 dwellings a hectare (UK Department of the Environment, Transport and the Regions, 2000).

10 A similar conclusion was reached from the Norfolk household survey. In response to a checklist question, 21.5 per cent indicated dissatisfaction with the availability of local jobs for school-leavers and 14.7 per cent for part-time jobs. There were higher levels of concern about police services (42.8 per cent), affordable housing (39.7 per cent), indoor meeting spaces for teenagers (38.9 per cent), public transport links to Norwich (33.3 per cent) and leisure and sports facilities (32.7 per cent).

> ... when Partnerships were being set up a few years ago, ... there was never a desperate need for economic regeneration, but that was partly because of the dormitory thing, you know, if you are 100 per cent dormitory, then economic vitality is not really important, as long as you've got a few local shops, because everybody is just sleeping here ... I am not conscious of any great effort to bring employment into Wymondham, we have had several designated industrial estates lying fallow for three or four years, well ten or fifteen years.

This is not to downplay the role of local government business grants, Norfolk's focus on strategic growth sectors (Norfolk County Council, 2001) and government initiatives to improve local skills and entrepreneurship. It does highlight that economic policies tend to be general rather than place specific. Without more positive support for the generation of employment in larger rural centres, it is doubtful that strong links will be fostered with surrounding villages. In addition to concerns about public transport services between villages and market towns, anxiety for the future is raised on two fronts. The first is that sustainable development principles will limit housing development in market towns that are experiencing slow employment growth, thus constraining the supply of affordable housing. The second is that sustainable development policies favour urban containment and the re-use of brownfield land, which could limit service centre expansion. In the case of Norfolk, the most developable brownfield land is in larger towns (Norfolk County Council, 2002a, p.31). With local authorities under pressure to utilize brownfield sites, this could limit future employment opportunities in service centres, thus reinforcing the orientation of rural towns and villages toward Norwich.

## Conclusion

From the detail presented above, the Norwich city-region supports the generalized depictions of land-use change trends that are presented in the literature, as well as policy emphases in the European Spatial Development Perspective. Land-use policies promoting the concentration of development during the 1970s and 1980s had mixed results. The concentration of development within Norwich's northern suburbs and selected growth settlements must be viewed in conjunction with continuing housing growth within smaller rural villages. During the 1990s the infiltration of sustainable development principles into planning policy guidance, combined with the European Spatial Development Perspective, strengthened planning commitments toward urban containment. This is exemplified in the Norwich city-region in the directing of public resources toward enhancing the inner-city, including greater attention to the redevelopment of brownfield land, and secondly, in the priority given to greenfield sites on the urban edge. Similar trends exist in other English city-regions, with brownfield redevelopment considered critical to the delivery of urban regeneration (UK Parliamentary Office of Science and Technology, 1998), while sustainable development principles question the importance given to greenbelt protection over other policy objectives (see, for example, Cambridgeshire County Council and Peterborough City Council, 2003).

The historical importance of Norwich as the regional centre for Norfolk, combined with more recent public policy efforts to protect and strengthen inner-city functions, has resulted in strong rural to urban population flows to access work, retail and leisure opportunities. This provides partial evidence that the ESDP's objective of strong rural-urban linkages is being realized. Yet, at the same time, it contradicts the ESDP objective of reducing travel volumes, particularly given the high rate of commuting by private vehicles. As adopted in the UK, sustainable development principles, rather than supporting the dispersal of economic activity, provide a strong argument for redirecting housing growth. In the case of Norfolk, to prevent rural housing developments from strengthening rural-urban commuting, priority is given to designating housing land closer to cities.

The importance of directing Norwich-related housing growth closer to built-up areas has a longer history, as symbolized by the introduction of the Norwich Policy Area in the mid-1970s. In addition to suburban expansion, strategic importance has been given to large-scale housing extensions in a ring of villages lying just beyond the urban fringe. Located near transport corridors into the city, the intention is that these hinterland settlements will be closely connected to Norwich by public transport, thus avoiding dispersed long distance commuting. Yet, as highlighted in Table 6.6, the rate of commuting to Norwich by private vehicles does not appear to be significantly different from more distant rural service towns. The high intensity of commuting between these hinterland settlements and Norwich has added to uncertainty over their function. Whilst environmental gains are associated with consolidating development in and around Norwich, the concentration of housing growth on the edge of existing settlements has not been adequately balanced with additional services. Community leaders have therefore criticized policy-makers for transforming existing rural villages into dormitory 'suburbs', and for failing to provide services. This is at odds with the ESDP's objective of safeguarding the quality of life of hinterland residents, and of the promotion of sustainable communities within more recent UK government policy (UK Office of the Deputy Prime Minister, 2004). Whilst a more extensive range of Section 106 contributions have been requested from developers in recent years, local needs are somewhat haphazardly balanced against city-region priorities. Looking to the future, one response is that housing growth will be more comprehensively master-planned either as urban fringe land release or as new towns (Cambridgeshire County Council and Peterborough City Council, 2003). Such developments will be larger, with more emphasis on Section 106 contributions for services and community infrastructure from developers. Whilst this style of land development will help the delivery of sustainable communities by ensuring that environmental and social benefits can be obtained, the realization of such developments will depend on their passage through the land-use planning system. If, for example, the negotiation of Section 106 contributions proves contentious, perhaps because developers are required to make funds available for wider city-region projects (including orbital roads), then the delivery of housing will be slowed. In this scenario demand for housing will spread outwards from cities, thus inflating house prices within city-regions and contradicting policies to limit commuting toward cities.

With land-use planning principles focused on directing future growth, less importance is given to how existing developments can be made more sustainable. This conclusion has particular relevance where rural villages experience declining service levels or are under the influence of urban housing markets. Where house prices have inflated beyond the purchasing capacity of local house-buyers, public policy has focused on the need to construct additional dwellings. But planning policies actively discourage housing building in hinterland villages, where heightened out-commuting is likely to result. For local authority officers responsible for social needs, this is a source of anxiety, because '... sustainability in planning terms focuses on reducing private transport use, rather than helping to maintain family and social networks' (South Norfolk District Council, 2002a, p.14). Whilst the Government's rural exception policy has been promoted as a way of meeting local housing needs, it does not appear to be conducive to the construction of many dwellings (Hoggart and Henderson, 2005). In addition to moral concerns expressed by planning officers about placing people in villages where services are limited, the demand for rural exceptions homes is precarious to calculate and is substantially less than reported in local housing need surveys.

Rather than providing arguments for improving the sustainability of rural villages, a key concern of this chapter is whether larger rural towns are able to meet the needs of households living in surrounding villages, as desired by current national policies. A particular concern for this chapter is the way sustainable development principles encourage housing growth to be directed toward urban centres, when the ability of rural towns to generate employment growth is constrained. In both UK Government policy and the ESDP, there is a lack of coherence over whether strengthening commuter links between market towns and nearby larger centres is problematic. On the one hand, the ESDP indicates that: 'The future prospects of the surrounding rural areas are also based on competitive towns and cities' (Commission of the European Communities, 1999, p.22). On the other hand, it acknowledges the need to strengthen '... small and medium-sized towns in rural areas as focal points for regional development' (p.25). For larger rural settlements, which increasingly depend on nearby urban centres, there is less clarity in terms of policy advice. In the case of Norfolk, the County Council indicates that the '... scale [of future housing in rural service towns] should be related closely to the range and quality of employment, services and facilities in each town' (Norfolk County Council, 2002a, p.31). With many larger rural towns experiencing difficulty in stimulating employment growth, recent policy reviews have proposed concentrating a greater proportion of Norfolk's future housing nearer to Norwich. Yet failing to tackle employment problems in rural towns will limit the viability of public transport links with surrounding villages, and direct the attention of rural residents towards Norwich. Restrictions on future housing development will also constrain the delivery of suitable housing for less mobile and lower income rural residents. This needs to be seen in the context of Government policy, which promotes higher density developments, with Section 106 agreements providing almost one half of the affordable housing units delivered nationally in 2003 (Monk et al., 2005), and with concern that the Housing Corporation's definition of 'rural' varies from other government bodies, including the Countryside Agency. Unless market towns can provide people from surrounding villages with desirable living opportunities, then pressure for additional growth in smaller settlements is likely to remain strong. Without more holistic policy, including policies to support a renaissance across England's larger rural towns, as in the past, attempts to strengthen the role of key rural settlements will be constrained.

# References

Banister, D. (1994) Reducing the need to travel through planning, *Town Planning Review*, 65, 349-354
Beckett, S. and Madgett, J. (1993) *The History of Tasburgh*, Everett, Dereham
Blacksell, M. and Gilg, A.W. (1981) *The Countryside: Planning and Change*, Allen and Unwin, London
Bramley, G. and Smart, G. (1995) *Rural Incomes and Housing Affordability*, Rural Development Commission, Salisbury
Bunnell, G. (1995) Planning gain in theory and practice – negotiation of agreements in Cambridgeshire, *Progress in Planning*, 44, 1-113
Cambridgeshire County Council and Peterborough City Council (2003) *Cambridgeshire and Peterborough Structure Plan*, Cambridge and Peterborough.
Campbell, H. Ellis, H., Henneberry, J. and Gladwell, C. (2000) Planning obligations, planning practice and land-use outcomes, *Environmental and Planning B: Planning and Design*, 27, 759-775
Catherine Bickmore Associates (2003) *The State and Potential of Agriculture in the Urban Fringe*, Countryside Agency, Wetherby
Chaney, P. and Sherwood, K. (2000) The resale of right to buy dwellings: a case study of migration and social change in rural England, *Journal of Rural Studies*, 16(1), 79-94
Chinery, C. (2000) Population, in T. Heaton (ed.) *Norfolk Century*, Eastern Daily Press, Norwich, 9-28
Cloke, P.J. (1979) *Key Settlements in Rural Areas*, Methuen, London
Cloke, P.J. (1989, ed.) *Rural Land-use Planning in Developed Nations*, Unwin Hyman
Cloke, P.J. and Thrift, N.J. (1990) Class and change in rural Britain, in T.K. Marsden, P.D. Lowe, and S.J. Whatmore (eds.) *Rural Restructuring: Global Processes and Their Responses*, David Fulton, London, 165-181
Cloke, P.J., Milbourne, P. and Thomas, C. (1994) *Lifestyles in Rural England*, Rural Development Commission, Salisbury
Commission of the European Communities (1999) *ESDP – European Spatial Development Perspective: Towards Balanced and Sustainable Development of the Territory of the European Union*, Office for Official Publications of the European Communities, Luxembourg
Connell, J. (1974) The metropolitan village: spatial and social processes in discontinuous suburbs, in J.H. Johnson (ed.) *Suburban Growth: Geographical Processes at the Edge of the Western City*, Wiley, London, 77-100
Corfield, P.J. (1994) The identity of a regional capital: Norwich since the eighteenth century, in P. Kooij and P. Pellenbarg (eds.) *Regional Capitals: Past, Present, Prospects*, Van Gorcum, Assen, 129-148
Cross, D.F.W. (1990) *Counterurbanization in England and Wales*, Avebury, Aldershot
David Simmonds Consultancy (2004) *Norwich Riverside Keys to Redevelopment: A Case Study*, British Urban Regeneration Association, hhtp://www.bura.org.uk

DTZ Pieda Consulting (1998) *The Nature of Demand for Housing in Rural Areas*, Report for the Department of the Environment, Transport and the Regions, Edinburgh

*Eastern Daily Press* (1990) Plans for a new village under attack, 22 March, 11

*Eastern Daily Press* (1990a) Council will press for rural housing, 27 November, 7

*Eastern Daily Press* (1996) Norfolk the poor relation of East Anglia, 13 November, Norwich Library Archive

Edwards, B., Goodwin, M. and Woods, M. (2003) Citzenship, community and participation in small towns: a case study of regeneration partnerships, in R. Imrie. and M. Raco (eds.) *New Labour, Community and Urban Policy*, The Policy Press, Bristol, 181-204.

Elson, M.J. (1986) *Green Belts: Conflict Mediation in the Urban Fringe*, Heinemann, London

*Evening News* (1988) Scheme 'attempt to buy' consent, 14 December, Norwich Library Archive

Faludi, A. (2003) Unfinished business: European spatial planning in the 2000s, *Town Planning Review*, 74, 121-140

Findlay, A.M., Stockdale, A., Findlay, A. and Short, D. (2001) Mobility as a driver of change in rural Britain: an analysis of the links between migration, commuting and travel to shop patterns, *International Journal of Population Geography*, 7, 1-15

Ford, T. (1999) Understanding population growth in the peri-urban region, *International Journal of Population Geography*, 5, 297-311.

Freestone, R. (1992) Sydney's green belt, *Australian Planner*, July, 70-77

Gallent, N. and Bell, P. (2000) Planning exceptions in rural England: past, present and future, *Planning Practice and Research*, 15, 375-384

Gilg, A.W. (1991) *Countryside Policies for the 1990s*, CAB International, Wallingford

Hall, P. (1989) *London 2001*, Unwin Hyman, London

Hall, P.G. (2002) *Urban and Regional Planning*, 4$^{th}$ edition, Routledge, London

Harvey, D.W. (1989) From managerialism to entrepreneurship: the transformation of urban governance in late capitalism, *Geografiska Annaler*, 71B, 3-17

Hedges, A. (1999) *Living in the Countryside: The Needs and Aspirations of Rural Populations*, Countryside Agency, Wetherby.

Hoggart, K. (1997) Home occupancy and rural housing problems in England, *Town Planning Review*, 68(4), 485-515

Hoggart, K. (2003) England, in N. Gallent, M. Shucksmith and M. Tewdwr-Jones (eds.) *Housing in the European Countryside: Rural Pressure and Policy in Western Europe*, Routledge, London, 153-167

Hoggart, K. and Henderson, S. (2005) Excluding exceptions: housing non-affordability and the oppression of environmental sustainability?, *Journal of Rural Studies*, 21, 181-196.

House Builders' Federation (1998) *Urban Life: Breaking down the barriers to brownfield development*, The House Builders' Federation, London

Ilbery, B. W. (1988) Agricultural change on the West Midlands urban fringe, *Tijdschrift voor Economische en Sociale Geografie*, 79(2), 108-121

Kearns, A. and Turok, I. (2003) *Sustainable Communities: Dimensions and Challenges*, ESRC Urban and Neighbourhood Studies Research Network, ODPM, London

Lawless, P. and Brown, F. (1986) *Urban Growth and Change in Britain*, Paul Chapman, London

Macnaughten, P. and Urry, J. (1998) *Contested Natures*, Sage, London

Marsden, T. K. (1995) Beyond agriculture? regulating the new rural spaces, *Journal of Rural Studies* 11(3), 285-296.

Marsden, T.K., Murdoch, J., Lowe, P.D., Munton, R.J.C. and Flynn, A. (1993) *Constructing the Countryside*, UCL Press, London

McNamara, P. (1984) *Restraint Policy in Action: Housing in Dacorum and North Hertfordshire*, Oxford Polytechnic Department of Town Planning Working Paper 77, Oxford
Monk, S. and Whitehead, C.M.E. (1996) Land supply and housing: a case study, *Housing Studies*, 11, 407-423
Monk, S., Crook, T., Lister, D., Rowley, S., Short, C. and Whitehead, C. (2005) *Land and Finance for Affordable Housing: The Complementary Roles of Social Housing Grants and the Provision of Affordable Housing Through the Planning System*, Joseph Rowntree Foundation, York.
Mormont, M. (1987) The emergence of rural struggles and their ideological effects, *International Journal of Urban & Regional Research*, 7, 559-578.
Munton, R.J.C. (1974) Farming on the urban fringe, in J.H. Johnston (ed.) *Suburban Growth*, John Wiley, London, 201-224
Munton, R.J.C. (1983) *London's Green Belt: Containment in Practice*, Allen and Unwin, London
Munton, R.J.C., Whatmore, S.J. and Marsden, T.K. (1987) Reconsidering urban-fringe agriculture: a longitudinal analysis of capital restructuring on farms in the metropolitan green belt, *Transactions of the Institute of British Geographers*, 13, 324-336
Murdoch, J. and Marsden, T.K. (1994) *Reconstituting Rurality: Class, Community and Power in the Development Process*, UCL Press, London
Norfolk County Council (1980) *Norfolk Structure Plan*, Norwich
Norfolk County Council (1993*) Norfolk Structure Plan*, Norwich
Norfolk County Council (1994) *Norfolk Structure Plan: The Norwich Policy Area to 2006: Background Paper: Environment, Consultation Draft*, Norwich
Norfolk County Council (1999) *Norfolk Structure Plan*, Norwich
Norfolk County Council (2000) *1996-based population projects for Norfolk*, Demographic Information Note 3/00, Norfolk County Council, Norwich
Norfolk County Council (2001) *Shaping the Future Strategy: The Economic Development Strategy for Norfolk 2001-10*, Norfolk County Council, Norwich
Norfolk County Council (2002) *First results from the 2001 census and mid-2001 population estimates for Norfolk*, Demographic Information Note 3/02, Norfolk County Council, Norwich
Norfolk County Council (2002a) *Norfolk Structure Plan Review Issues Report: Looking Towards 2025*, Norfolk County Council, Norfolk
Norfolk County Council (2002b) *Norfolk County Council Transport Guide*, Norfolk County Council, Norwich.
Norfolk County Council (2003) *Norfolk Structure Plan (1999) Monitoring Statement*, Norfolk County Council, Norwich, http://www.norfolk.gov.uk
Norfolk County Council (2004) *Local Transport Plan: Appendix B: Significant developments for our 2nd Local Transport Plan*, http://www.norfolk.gov.uk/transport/transportplanning/ltp/pdf/fourthreport/Appendix%20B%20-%20LTPReport.pdf
Norfolk County Council (2004a) *2001 Census Topic Report on Resident, Workplace and Daytime Populations*, Norfolk County Council, Norwich, www.norfolk.gov.uk/demography
Norfolk County Council (2005) Historic Census Population Figures Parish by Parish Back to 1891, Norfolk County Council, Norwich http://www.norfolk.gov.uk/council/statistics/demography/excel/historic_parish.xls
*Norwich Mercury* (1965) Wymondham booming, 26 February, Norwich Library Archive
*Norwich Mercury* (1971) Wymondham 'not city dormitory', 3 February, Norwich Library Archive

Ostendorf, W. (2001) New towns and compact cities: urban planning in the Netherlands between state and market, in H. Andersson, G. Jorgensen, D. Joye and W. Ostendorf (eds.) *Change and Stability in Urban Europe: Form, Quality and Governance*, Ashgate, Aldershot, 177-192

Owens, S.E. and Cowell, R. (2002) *Land and Limits: Interpreting Sustainability in the Planning Process*, Routledge, London

Pacione, M. (1990) Private profit and public interest in the residential development process: a case study in the urban fringe, *Journal of Rural Studies*, 6, 103-116.

Pacione, J. (2004) Where will the people go? assessing the new settlement option for the United Kingdom, *Progress in Planning*, 62(2), 73-129

Pahl, R.E. (1965) *Urbs in Rure: The Metropolitan Fringe in Hertfordshire*, London School of Economics Geographical Paper 2, London

Perkins, M, Day, H. and Heaton, T. (2000) Industry and Commerce, in. T. Heaton, (ed.) *Norfolk Century*, Eastern Daily Press, Norwich, 162-182

Philips, D.R. and Williams, A.M. (1984) *Rural Britain: A Social Geography*, Blackwell, Oxford

Popper, F.J. (1985) The environmentalist and the LULU, *Environment*, 27(2), 7-11, 37-40.

Poynton, L. (1988) £1m pledge for leisure facilities, *Evening News* 14 December, Norwich Library Archive

Sawyer, A. (2002) *Diss: A Market Town Planning for a Prosperous Future*, Diss Development Partnership, Diss

Sawyer A. (2002a) Harleston: A Special Market Town with Historic Heritage and a Bright Future, Harleston Development Partnership, Harleston

Shaw, J.M. (1980) *Services in Rural Norfolk 1950-1980: A Survey of the Changing Pattern of Services in Rural Norfolk Over the Last Thirty Years*, Norfolk County Council, Norwich

Shucksmith, M., Watkins, L. and Henderson, M. (1993) Attitudes and policies towards residential development in the Scottish countryside, *Journal of Rural Studies*, 9, 243-255

South Norfolk District Council (2002) Planning Applications and Other Development Control Matters, Main Planning Committee Agenda, 10 April, http://www.south-nofolk.gov.uk/south-norfolk/council.nsf/pages/planmins100402a.html

South Norfolk District Council (2002a) *Housing Strategy 2002/03*, South Norfolk District Council, Long Stratton

South Norfolk District Council (2003) *South Norfolk Local Plan*, South Norfolk District Council, Long Stratton, http://www.south-norfolk.gov.uk

Tewdwr-Jones, M. (1997) Green belts or green wedges for Wales: a flexible approach to planning in the urban periphery, *Regional Studies* 31, 73-77

*The Guardian*, 2004, Rural areas 'need new homes', 31 March 2004, 8

UK Department of the Environment (1988) *Housing in Rural Areas: Village Housing and New Villages*, HMSO, London

UK Department of the Environment, Transport and the Regions (2000) *Planning Policy Guidance Note 3: Housing*, Stationery Office, London

UK Department of the Environment, Transport and the Regions (2000b) *Our Towns and Cities: The Future: Delivering an Urban Renaissance*, HMSO, London

UK Department of the Environment, Transport and the Regions (2000c) *Our Countryside: The Future – A Fair Deal for Rural England*. HMSO, London

UK Department of Transport (1994) *Planning Policy Guidance Note 13: A Guide to Better Practice – Reducing the Need to Travel Through Land-Use and Transport Planning*, HMSO, London

UK Office for National Statistics (2003) *Census 2001: Key Statistics for Local Authorities in England and Wales*, HMSO, Norwich
UK Office for National Statistics (2003a) *Job density 2001*, Regional Trends 38, http://www.statistics.gov.uk/STATBASE/ssdataset.asp?vlnk=7709
UK Office of Population Censuses and Surveys (1994) *1991 Census Key Statistics for Local Authorities*, HMSO, London
UK Office of the Deputy Prime Minister (2004) Planning Policy Statement: Delivering Sustainable Development, http://www.odpm.gov.uk/stellent/groups/odpm_planning/documents/pdf/odpm_plan_pdf_035506.pdf
UK Office of the Deputy Prime Minister (2005) *Delivering Sustainable Communities: The Role of Local Authorities in the Delivery of New Quality Housing*, ODPM, London
UK Parliamentary Office of Science and Technology (1998) *A Brown and Pleasant Land: Household Growth and Brownfield Sites*, Report 117, Parliamentary Office of Science and Technology, London
Valler, D. (1996) Locality, local economic strategy and private sector involvement: case studies in Norwich and Cardiff, *Political Geography*, 15, 383-403
Wannop, U. (1999) New towns, in B. Cullingworth (ed.) *British Planning: 50 Years of Urban and Regional Policy*, Athlone, London, 213-231
Warnes, A.M. (1991) London's population trends: metropolitan area or megalopolis?, in K. Hoggart and D.R. Green (eds.) *London: A New Metropolitan Geography*, Edward Arnold, London, 156-175
While, A., Jonas, A.E.G. and Gibbs, D.C. (2004) Unblocking the city? Growth pressures, collective provision, and the search for new spaces of governance in Greater Cambridge, *Environment and Planning, A*36, 279-304
World Commission on Environment and Development (1987) Our *Common Future*, New York
Yarwood, R. (2002) Parish councils, partnership and governance: the development of 'exceptions' housing in the Malvern Hills District, England, *Journal of Rural Studies*, 18, 275-291

Chapter 7

# Convergence and Divergence in European City Hinterlands: A Cross-National Comparison

Keith Hoggart

**Introduction**

Dimensions of the European Spatial Development Policy (ESDP) have informed the research that has been presented in the chapters of this book. Fundamental to the ESDP is a vision of the future that emphasizes the importance of spatial planning and which promotes more integrated actions across policy fields, as compared with the sectoral focus that dominated EU policy-making in the past (e.g. Faludi and Waterhout, 2002). Within the fabric of ESDP conceptualizations, polycentric development is favoured, in which three scales are acknowledged: the inter-regional, for which the creation or encouragement of multiple growth regions is seen as desirable; the intra-regional, which envisages sustained socio-economic dynamism for a number of city-regions; and, the city-region, in which the promotion of multiple growth points and the integration of urban and rural are seen as fundamental (e.g. Tewdwr-Jones and Williams, 2001, p.38). For this book, the key geographical scale has been the city-region. Within a city-region framework, critical questions have been asked about the construction and maintenance of urban-rural partnerships (Commission of the European Communities, 1999, p.25). The critical ESDP assumption is that rural and urban are not distinctive. Rather they are integrated, in a mutually dependent and reinforcing interchange, which ensures that urban problems are also rural problems, and vice versa. Hence, rural areas around cities are in a two-way dynamic, so cities and their hinterlands require integrated spatial development strategies. Within the ESDP, it follows that city-regions are not 'black boxes', nor are hinterland zones central city appendages. More accurately, city-regions are conceived as comprised of overlapping rural-urban influences. As the incidence and intensity of change forces varies in urban as well as in rural realms, so does the nature of their interactions, with the success of efforts to stimulate socio-economic conditions '... very dependent on local conditions' (Commission of the European Communities, 1999, p.22). But which local conditions are critical?

Here the Commission is explicit in identifying critical issues for sustainable development. These all focus on aspects of the quality of life of city and hinterland residents but for this book the most central are those that bear on the relationship between employment, housing and services. This is not to discount the importance of ecosystem management, nor of issues like controlling the physical expansion of built-up areas. These are both intricately entwined with the issues explored here. But their in-depth investigation requires more space than can be allocated in this book. Moreover, central tenets of the ESDP create tensions that merit in-depth exploration of interchanges between employment, housing and services. This is well illustrated by the UK, which is one nation that has adopted ESDP principles in noteworthy ways (Tewdwr-Jones and Williams, 2001; Faludi, 2003). Capturing central ESDP principles of reducing urban sprawl, promoting functional and social heterogeneity, improving transport accessibility, while providing for environmental enhancements that conserve natural and cultural heritage (Commission of the European Communities, 1999), contemporary UK policies draw on environmental sustainability principles to promote the regeneration of urban centres, the protection of the countryside, and improved access to jobs and services. At its core, this is achieved through the land-use planning system, which directs new dwellings toward existing built-up areas. Only once brownfield possibilities have been discounted are suburban, then hinterland small town sites supposed to be explored, with development in smaller settlements discouraged:

> In identifying sites to be allocated for housing in local plans and UDPs, local planning authorities should follow a search sequence, starting with the re-use of previously-developed land and buildings within urban areas identified by the urban housing capacity study, then urban extensions, and finally new development around nodes in good public transport corridors. (UK Department of the Environment, Transport and the Regions, 2003, p.10)

The acceptance of some growth in secondary centres within city-regions is based on the notion that housing expansion can be balanced with employment and services. There is an aversion in the thought processes underlying these objectives to the emergence of dormitory settlements surrounding regional cities. In this regard, a fundamental goal is to minimize car-based commuting. A key underpinning for this goal is that, if a strategic balance can be struck between employment, housing and services, people should travel less, with private car usage declining, and harmful carbon dioxide emissions lessened (UK Department of the Environment, Transport and the Regions, 2003a). Yet the validity of this assumption has been questioned, with hopes that compact cities and job-home proximity will reduce travel failing to find convincing empirical support (e.g. Diepen, 2000; Breheny, 2004).

Further troubling for the implementation of ESDP objectives is the manner in which their execution throws up difficulties for other social objectives. With its longstanding adherence to compact city philosophies, the Netherlands provides the starkest exposition of these concerns. In particular, as the Dutch Government's desire for compact cities materialized in new construction, the relative attractiveness of city and countryside altered, with wealthier households, who are better able to choose where they live, able to select more exclusive rural abodes. Rising land prices owing to a scarcity of housing land in or adjacent to cities, alongside the reluctance of private developers to build in a manner the Dutch Government sought, led to suburbanization, with a consequent geographical segregation of income groups (e.g. Ostendorf, 2001). This bodes ill for expectations that the ESDP will provide social and functional heterogeneity. Yet precisely this vision is being built into national policy directives, as illustrated by the UK Government's *Planning Policy Guidance 3: Housing*, which states that:

> The Government believes that it is important to help create mixed and inclusive communities, which offer a choice of housing and lifestyle. It does not accept that different types of housing and tenures make bad neighbours. Local Planning authorities should encourage the development of mixed and balanced communities: they should ensure that new housing developments help to secure a better social mix by avoiding the creation of large areas of housing of similar characteristics. (UK Department of the Environment, Transport and the Regions, 2003, p.5)

**Table 7.1    Welfare Regimes and Housing Rental Systems**

| EU Member State | Welfare regime | Rental market type |
| --- | --- | --- |
| Sweden | Social Democratic [SD] | Unitary |
| Denmark | Social Democratic [SD] | Unitary |
| Finland | Social Democratic [SD] | Unitary |
| Netherlands | Intermediate [C - SD] | Unitary |
| Germany | Corporatist [C] | Unitary |
| Austria | Corporatist [C] | Unitary |
| France | Intermediate [SD - LW] | Unitary |
| Ireland | Liberal Welfare [LW] | Dual |
| UK | Liberal Welfare [LW] | Dual |
| Italy | Rudimentary Welfare [RD] | Dual |
| Greece | Rudimentary Welfare [RD] | Dual |
| Portugal | Rudimentary Welfare [RD] | Dual |
| Spain | Rudimentary Welfare [RD] | Dual |

Source:    Ball and Grilli (1997, Table 2.1)
Note:       In Social Democratic [SD] states the dominant force is universalism and decommodification, extended to all social classes, to produce 'one nation' welfare provision. Corporatist [C] states reinforce the rights of social classes and professions, and replace market provision to do so. Liberal Welfare [LW] states provide little more than a safety net for those on low-incomes. Rudimentary Welfare [RW] states are akin to Liberal Welfare, but with weaker provision. In dualist rental systems, government policies are a prime factor in preferences for owner-occupation (Kemeny, 1995), whereas in unitary systems housing tenure is treated comparably, with owner occupation more from choice than 'push' factors.

The question this prompts for European nations is whether there is a transposition of different political cultures into similar or dissimilar outcomes, as regards the satisfaction of ESDP aims. The key point here, which Esping-Andersen (1990) and others have exemplified, are the different political philosophies that lie at the heart of European societies. The importance of this divergence is seen in different ways across activity spheres. Thus, Table 7.1 takes the kind of general social welfare typology that Esping-Andersen has produced and compares it with Kemeny's (1995) characterization of housing markets. Critically for this book, the four nations that form the core for analysis have dissimilar social welfare regimes and divergent housing market systems. As such, we might expect dissimilar responses, from civil society and the state, to issues like social exclusion, as well as quite distinctive housing market processes. As a result of these considerations, plus differences in the nature of land-use planning across the continent, there are reasonable grounds for expecting disparities in the nature of change within city hinterlands. How far this has proved to be the case is explored here by considering the key messages of earlier chapters in the context of ESDP aims for functional heterogeneity, social inclusion and access to services.

**Functional Heterogeneity**

In truth, when we look at functional distinctions within the city-regions investigated, it is significant that, in terms of relations between city and hinterland, there is little to distinguish experiences across the four countries. If we take Annecy as an exemplar, here we find a concentration of jobs within the City itself, with secondary centres having a very subsidiary role (Chapter Four). For sure, subsidiary or second level centres are important loci for employment, but the core characterization of the city-region sees the central city as the focus of most jobs. Precisely the same point can be made for Norwich, where regional administrators report that the city represents the primary attraction force for outside investors who consider locating in the city-region, while those seeking to promote the economies of surrounding market towns lament the passage of their urban centre into dormitory town status (Chapter Six). The message that the central city is the core employment node also characterizes the Granada and Munich city-regions. In the case of Munich, which is much larger than the other cities investigated, there are noteworthy foci for employment in the suburbs and beyond; as seen for mini-concentrations in secondary centres (like county capitals) and near the airport (Chapter Three). Yet, the main locus of paid-work is still in the City and its suburbs, with few job openings in rural and small town hinterland areas. With population growth generally occurring more rapidly in hinterland settlements than in core (or secondary) cities, a growing separation of home and workplace is occurring. This has led to longer commuting times for hinterland residents to access jobs, with time spent in the core city raising the attractiveness of this centre as a venue for retail and leisure activities. The same pattern characterizes Granada, where many residents are described as being committed to the municipality of their home largely for sleeping and weekend activities (Chapter Five, p.105). As with the other activities explored here, the expansion of hinterland settlements around Granada has not been accompanied by significant job decentralization. Even in

Norwich, where the land-use planning system imposes tighter constraints on residential expansion, while strategic planning goals are to produce greater correspondence between the employment and housing (Norfolk County Council, 1999), there has been disproportionate population growth in rural villages (Norfolk County Council, 1999, p.63). This has occurred not only without accompanying local job formation, but often with a loss of employment opportunities in market towns and rural areas (due to agricultural decline, the loss of agro-food industries and the decline of light manufacturing), with the result that there has been an increasing dependence on Norwich and its suburbs for jobs, along with growth in work-related commuting to that centre. Yet this does not result in significantly poorer levels of employment in the rural hinterlands (e.g. Henderson and Hoggart, 2003), as the lack of local jobs is responded to by commuting to the core city. In this regard our case study cities have obvious similarities, with Valence as a partial exception, as this city stands in the midst of a triangle of job centres, while El Ejido's reliance on agriculture as the driving force behind its economic expansion makes a somewhat different mark, as the fruit and vegetable economy of the area favours more proximity between work and residence.

But in terms of understanding the character of change in hinterland zones, it is not sufficient to acknowledge that jobs are primarily concentrated in the central cities (and their suburbs), for a key issue in all nations is the mismatch between the maintenance of the central city as an employment node and the vibrancy of residential growth elsewhere. In the case of Granada and Munich, the high price of housing in the central city (and suburbs) encourages residents to look for cheaper options in hinterland settlements, with lower prices often favouring smaller, sometimes scattered settlement structures. In Granada this cost incentive is heightened by the inadequacy of available property in the central city's housing market, which still has an excess of poor quality homes, while also experiencing shortages in social housing and limited land for new-build. The limited size and facilities of old apartments and houses in the City combine to draw households toward larger properties with land. Such properties can be purchased or rented cheaply in hinterland towns and villages compared to smaller, less adequately provided dwellings in the central city. In all nations, a shortfall of cheap housing in central cities places increasing demands on hinterland settlements, given that they generally offer less costly accommodation. England offers something of an exception to this, with the attraction of rural homes increasingly resulting in village properties carrying a price premium (e.g. Countryside Agency, 2004), with resistance to homes in rural areas and a reluctance of institutional providers to respond to rural housing needs when those in cities are higher, resulting in fewer opportunities for low income households to secure rural homes (e.g. Hoggart and Henderson, 2005).

Although expressed in varying form – from the addition of what can be micro-units grafted onto existing micro-settlements, as in the Munich case, to the shotgun dispersal of new-build around Granada – one feature that separates experiences in France, Germany and Spain from that of England is the limited range of English settlements which are allowed to see noteworthy new dwelling increments. In Germany new-build is limited to existing settlements, yet construction can occur in a relatively scattered manner across a variety of hamlets or small villages. This arises because regional authorities prescribe that 'organic growth' should be dominant in smaller municipalities, with the absence of a local plan leading to any settlement being acceptable as a site for development. In contrast, new dwellings in England are increasingly restricted to a short list of settlements (predominantly larger ones), save for a meagre number of new homes that are allowed as 'exceptions' to current planning policy (Hoggart and Henderson, 2005). Hence, while there is considerable demand for living in rural areas (Countryside Commission, 1997), much of this demand is not satisfied owing to land-use planning restrictions. Demand for limited rural housing opportunities pushes prices up in England, so they are now often comparable with urban centres (and given wages are often lower in rural zones, the share of income devoted to housing is higher than in cities; e.g. East of England Regional Housing Forum, 2003). By contrast, in France, Germany and Spain housing expansion in rural and small town hinterland settlements owes much to the relative cheapness of housing, and in some cases to the less satisfactory standard of homes in central cities. Hence, in these three nations, centrifugal forces are at play, which push households away from the city. By contrast, for the English, the countryside draws in-migrants more forcefully, with this basic difference embodied in fundamental national values concerning the countryside (e.g. Hoggart *et al.*, 1995). Yet the attraction of the countryside is growing in many European nations (e.g. Paniagua, 2001); with national differences not paralleling national typologies like those in Table 7.1.

In summary, hinterland territories are drifting toward a functional role within city-regions that are increasingly residential (albeit sometimes in a second home or tourism context). The fact that El Ejido in Spain is the partial exception to this is poignant (Chapter Five), as the basis of this city-region's economic advancement is intensive fruit and vegetable production. The absence of agriculture as a focus of commentary in the discussion so far is instructive here, for, in terms of the functional contribution of hinterland territories to their city-regions, agriculture is on the slide. The huge income differential garnered from harvesting land for buildings rather than for crops or animals is such that farmers close to cities see massive financial rewards accruing from selling land for residential, leisure or other services. This does not mean farmers do this willing, nor that they are keen to do so. But operating close to the city raises real problems for farmers, as incomers reject the noise, smells and traffic associated with farm operations, while also causing damage to farm activities. In a context in which European agriculture, or at least many parts of it, are experiencing financial stress, the capacity to sell parcels of land offers a life-line, with higher land prices closer to the city enabling farmland to be sold in order to purchase larger holdings away from the city (Chapter Three), while those who leave farming release land for purchase or rent by neighbours who want to increase the size of their operations (Chapter Four). In any event, agriculture is declining as a fundamental element in city-hinterland functional relationships. Although by no means universal in its occurrence, where tourism potential and global marketing opportunities give local areas sales advantages, these make much more weighty impacts on the dynamics of hinterland change. Note, for example, how, around Granada, inadequacies in the road system, as in Güéjar-Sierra, pose the real limitation on tourism inspired housing growth, not job shortages, prices or land availability.

A further issue about trends in hinterland zones, is that both tourism and farm intensification offer the prospect of less dependency on the central city, by offering more localized job opportunities. On the surface this might seem to bode well for the ESDP's environmental sustainability agenda. Yet the reality is that these localized economic opportunities also threaten local environments. This is apparent in Spain, where both agricultural intensification in El Ejido (e.g. Tout, 1990) and agricultural decline, as in Granada (e.g. Mata Porras, 1995), threaten environmental quality, especially as possibilities of securing higher paid jobs in or in cities encourages land abandonment, with attitudes toward infusing land tenure with greater environmental awareness being slow to develop in Spain (Paniagua, 2001).

What all this implies, in terms of ESDP principles, is that current tendencies do not offer a comforting message. Rather than promising hinterland functional heterogeneity and improved city-region environmental sustainability, current trends point toward more homogeneity and increased travel distances for work, leisure and services.

## Social Inclusion

In terms of the social consequences of these outcomes, pressures (and desires) to find homes away from city cores, alongside an ongoing concentration of jobs and services in the city core, are generating price premiums for homes that are closer to the city. One result is that lower income families are better able to afford homes at a greater distance from the city (e.g. Morrison, 2003), which means they are more likely to incur higher transport costs or might even find it difficult to sustain access to employment positions, given travel costs and difficulties. Housing affordability is a significant issue for rural residents who wish to see their families continuing to live in hinterland areas that are zones of increasing inward migration. A key reason why this is a problem, across all the countries investigated, is because in-migrants tend to be relatively better-paid middle class and professional workers. Even if relatively low pay is the reason why some families feel they are driven toward more distant country homes, the incomes of these families are commonly higher than those of existing local residents, which creates local housing pressures.

The literature suggests that demand for rural homes is particularly noteworthy amongst younger middle class families who see hinterland destinations as desirable because they offer supposedly good environments to raise children (e.g. Murdoch, 1995). In the investigations presented in this volume, there is some evidence of this tendency (e.g. Chapter Four), but it is also clear that the population that flows into hinterland villages and towns is more heterogeneous than simplified idealizations of idyllic countryside lifestyles suggest. Within these middle class migrant streams, for example, we find increasing numbers of adults arriving without children (e.g. Hoggart, 2000; Heitkamp 2002; Ismaier 2002). Hence, there are more general forces at play, with each of France, Germany and Spain recording high demand for living in individual housing units (or at most in dwellings that contribute two homes), with this housing type closely identified with rural living. Indeed, while the looseness of land-use planning regulations in Germany means that multi-occupancy dwellings are relatively easy to provide if these are conversions of large farm barns or the like (Chapter Three), in general proposals to build or create multi-unit dwellings have met with resistance from local residents and recent arrivals. Clearly, the absence of multi-occupancy and 'terraced' housing options is likely to restrict options for those on lower incomes.

This is in a context in which rising rural land (and housing) prices are a cause for concern. In the case of France, it is not simply the impression of urban spill-over effects that is critical, for there are also cross-border forces at play, with Swiss nationals seeking housing and/or company outlets within border regions (the latter to secure a location within the European Union). This overlaying of urban-centred forces is also evident in Spain, where tourism is important, or, as in the case of El Ejido, where agricultural intensification has adjusted the urban-centred land price gradient. In effect, the spatial expression of the land price gradient is 'distorted' by having more driving forces in price determination. But whatever the generating impulses, the consequences of price increases are significant. In Germany, for example, while at one level land price hikes are resulting in smaller plot sizes, at another they are pushing the commuter-belt outwards from the central city, so increasing commuter distances and threatening environmental sustainability (see also Morrison, 2003). In France and England, in particular, rising prices in rural hinterland settlements are accompanied by a lack of social housing provision, so favouring the intensification of social differences between urban and rural areas. The precise form this is taking varies. In France and Germany, higher prices result in younger hinterland populations, or in middle class and professional households on relatively lower incomes. Of course, due to a lack of social housing, those with the lowest salaries find it difficult to secure a home in such locations, while lower incomes are linked to residences being farther from the central city. Also of note is the lack of rental options in hinterland settlements. In France and Germany this is especially notable, with developers finding the economics of private rental housing means that the hinterland construction of rental units yields insufficient profits.

Although the availability of social housing units is not considered a problem in the German case, what is an issue is the availability of affordable housing for local people. This statement reflects a different picture of what social housing is in Germany, with European nations having different approaches to addressing housing difficulties for those with relatively low incomes (e.g. Harloe, 1995). In England, for example, housing affordability difficulties are not necessarily associated with higher house prices but can result from relatively low incomes, with the corollary that those with quite high incomes on a national scale can still find themselves priced out of local housing markets (e.g. East of England Housing Forum, 2003). Hence, there have been moves to address so-called 'key worker' housing difficulties (such as teachers, nurses and the like), which arise when national salary scales are insufficient to meet accommodation costs in particular areas (e.g. Local Government Association, 2002; Morrison, 2003). Viewed under the broad heading of 'social housing' (in that subsidies are required to make housing affordable), the German situation is comparable to that in France and England, with the *Einheimischenmodell* policy reflecting efforts by municipalities to make housing available to local residents by enabling land to be purchased at subsidized prices in new residential areas (Chapter Three). Even in Spain, where land prices have traditionally been lower, there is a growing problem of housing affordability in hinterland areas subject to intense urban pressures, as in Monachil near Granada, where people see house prices as too high for their children to afford (Chapter Five). In all four nations, therefore, there are growing problems of affordability associated with housing expansion in city hinterlands, with limited resources to assist those who experience the negative consequences of land/ housing price increments.

As well as entrenching tendencies toward social segregation, current trends in hinterland zones are raising further social concerns, as seen in worries about the changing nature of local societies. In France, for example, confrontation is identified between rural and urban styles of life, that at times make themselves manifest in conflict, as over proposals for affordable housing (Chapter Four). In Germany newcomers are claimed not to integrate with longer term residents, with conflicts arising in part over demands for new services by incomers, especially given the cost of such services, which bear on the tax burdens of longstanding residents with lower incomes (Chapter Three). In addition, in-migrants are found to reject existing social organizations. Whether this is due to a dislike for 'traditional' organizations or such organizations do not accommodate newcomer needs, requires in-depth investigation. What is clear is that the wide range of clubs in most villages (sports clubs, shooting clubs, costume, etc) are demarcated by an absence of newcomer involvement. In Spain also there is tension, although this is not significant in areas with the most intense urban pressures e.g. (Monachil). Rather it is in La Alpujarra, where tension had arisen over newcomer demands for improvements in infrastructure, with long-time residents concerned about the costs of such investments, given low local incomes (Chapter Five). In England, as in Monachil, concern rather than conflict captures the sentiments of long-time residents toward newcomers. Here there is concern about the future quality of life of family members, who might have to move to obtain housing, given that incomers put pressure on local house prices, although there is recognition that the centrality of Norwich for services and employment is lessening the importance of local areas in people's lives (Chapter Six). This is seen to result in less interaction between neighbours, while the fact that services are in decline in hinterland villages is placing increasing pressure on the voluntary sector to provide more services and community support, with changes in the social composition of villages and in degrees of neighbourliness raising fears about the continued viability of voluntary efforts.

Linked to these concerns are fears about the erosion of cultural norms. It is notable, for example, that in all the case studies, except El Ejido, where intensive agriculture has developed, there has been a shift away from farm employment, which has consequences for the cultural base of villages. In France, for example, a loss of identity is associated with the transformation of farmland into urban uses, with buildings breaking up open space and splitting land holdings, so a fragmentation of farm operations has occurred. This has not only been seen in economic or landscape terms, but has in cultural loss, as a 30 per cent drop in farms over the last 12 years has been associated with a fall in farmer representation on rural community organizations. In Germany likewise infill housing is associated with a loss of village identity, as in-migrants bring new values to an area, and higher residential densities induce landscape change. In effect, 'urban lifestyles' and demands for urban services (as in France and Spain) challenge previously dominant cultural norms (e.g. over demand for more and different services).

It follows from the above that social heterogeneity is in a manner of speaking taking place. However, this is not the social diversity that is lauded in the ESDP, with those of different background engaged in social mixing. Rather it is a heterogeneity that results from the invasion of a new population that is in the process of re-writing the cultural norms and social characteristics of hinterland populations. Across city-regions, there is a trend toward the segregation of social groups, with hinterland zones taking a higher proportion of young households, who are not amongst those with the highest wages but are not the poorest paid. For the latter group the rising cost of accommodation is becoming an issue in all countries, although this is most marked in England, where land-use planning restrictions are tightest (Chapter Six). This provokes thoughts on the impact of land-use planning restraints, especially given suggestions for the USA that housing opportunities for the less advantaged are better in settlement structures that favour low-density urban sprawl (e.g. Kahn, 2001). In this context it is perhaps surprising that there is not more conflict arising between newcomers and longstanding residents. One possible reason for subdued tensions is agreement amongst incomers and existing residents alike about key development issues (such as multi-unit residential blocs). Another is that, although fearing aspects of change, there is some sympathy for the changes newcomers want (such as improved services). A third possibility is that people recognize that part of the change that is occurring is not due to newcomers as such, but represents broader changes in society (as with a weakening of local social ties). Whatever the cause, the case studies show that social tension should not be ignored but is not at the forefront of concerns over change in hinterland settlements. Viewed from a city-region perspective, a more pervasive social adjustment is the geographical realignment of socio-economic and demographic groups, which are producing a distinctive social profile in hinterland settlements.

**Access to Services**

This social differentiation is seeing cities retain much of the employment capacity of city-regions, even if rural tourism and activities unrelated to the urban hierarchy disrupt clean lines of distinction. This geographical pattern does not fit comfortably with the ESDP's vision of spatial development. The spirit of the ESDP embodies the notion of polycentric development – not just in terms of the positioning of city-regions within a national space-economy (although this has been subject to critical review over the positioning of rural areas; Richardson, 2000), but also in requiring '... partnership between towns and cities of every size and their surrounding countryside' (Commission of the European Communities, 1999, p.25). This multi-nodal perspective envisages a variety of centres offering employment and services for surrounding areas, even within the same city-region. But while deconcentration of investment from the central core is occurring, this is limited in effect, except where there are 'special' local conditions, such as those that inspire rural tourism. In the Munich case, for example, there has been considerable development close to the new airport, but this is not a model for heterogeneity in the distribution of economic activities within the city-region, unless equally powerful attractions are created elsewhere in the hinterland, for central cities are still dominant attractions for rapidly growing economic sectors that are not tied to specific places due to natural resource considerations (e.g. Bennett *et al.*, 1999). Offering an alternative model are the special circumstances of natural advantages (even if these take human ingenuity to realize their potential). For example, the driving force behind job increases outside Granada City has been tourism, although this has led to service provision that is not oriented toward the local population, but toward visitors. As regards services designed for local needs, in each of England, France and Spain, service and employment concentrations in central cities result in difficulties for smaller centres breaking away from a dormitory function, even if there has been some decentralized concentration, as in Annecy. In some cases, as with the UK, focused service provision is in line with government policy, with sustainable development the goal that promotes compact development.

On service provision, concentration (and deconcentration) pressures have clear implications for the ESDP's emphasis on urban-rural partnerships. Whatever the service type, service distributions cannot be divorced from infrastructural provision, which is a central element in ESDP thinking on sustainability. Put simply, if services are located in places that increase travel times to access them, this has implications for the environment as well as for social exclusion (as groups like the elderly, etc., will have greater difficulty accessing services). But service distribution issues are not just about numbers but also about types. In the Spanish case, for example, services growth in the Granada hinterland has been more notable for catering for tourists than for local residents (as in La Alpujarra, where services for local people are poor compared with those for tourists). Then again, the inflow of new residents has improved services, with those related to education, health, culture and transport increasing as the population has grown. In addition, in-migrants have prompted (or even set-up) new types of services in hinterland settlements, as with English-language nurseries in Barrio de la Vega, which have proved popular. However, as with kindergartens in German villages, while such services improve the spectrum of local facilities, demand for them is neither universal nor necessarily unaccompanied by costs to the municipal purse. Moreover, many still look to the central city to satisfy their needs, with many hinterland residents relying on the central city for a wide variety of services (as in Spain and England; also see Guy, 1991). Whether this is largely associated with commuters is an issue that required further exploration, for, as the Norwich case study shows, amongst many who live in villages around cities there is dissatisfaction over the expectation that they will satisfy their service needs in the central city (Chapter Six). With access issues having socially differential impacts, the relative absence of public transport in hinterland zones clearly links service issues to questions of social exclusion.

What should also be noted is the potentially debilitating impact of the current geography of new-build within hinterland zones on both landscape attractiveness and the economics of infrastructure provision. Near Granada, for example, we find extensive building on protected areas like the 'Vega', with weak municipal control over development processes, for only a few municipalities have a local strategic plan. But even if they do have such a plan, there are incentives to ignore it. A key reason for this is that the administrative process for new-build is slow in Spain, with examples recorded by the Spanish research team of construction projects taking up to 8-10 years from land purchase to completion and sale. Although this might be exceptional, a slow administrative process favours avoidance of local land-use stipulations, with some municipalities consequently not bothering to formulate a land-use plan. The resulting scatter of buildings is somewhat different from the situation in Germany, where new-build is at least restricted to existing settlements; although quite small hamlets are qualifying settlement in some areas. More in line with the Spanish situation, a good summary for land-use change in French rural hinterlands is a shift from a rural-urban dichotomy, where rural and urban are distinct land-uses, to a rural-urban continuum, where a mixing of rural and urban land-uses is prevalent. Yet, while tighter land-use regulations in the English countryside might point toward less landscape desecration and more economically viable infrastructure provision, English land-use planning outcomes do not fit comfortably with the ESDP goal of securing enhanced environmental sustainability, improved social inclusion and providing better services access.

## Concluding Thoughts

In meeting the objectives of the ESDP at the city-region level, the prescription provided points to the need: (a) to pursue the concept of 'the compact city' in order to control urban expansion; (b) to reconstruct derelict areas, provide appropriate access to basic services and facilities for all citizens and offer more open (green) space; (c) to adopt an integrated approach, 'with closed cycles' for natural resources, energy and waste, for effective environmental management; and, (d) to promote spatial policy that focuses land-uses and develops an integrated transport planning system so as to lessen car usage (Commission of the European Communities, 1999, pp.22-23). This chapter has explored the conclusions of the empirical analyses undertaken for each of England, France, Germany and Spain, with a view to identifying similarities and dissimilarities in change processes within city hinterlands, in order to cast light on these fundamental goals. While not all of these issues have been explored equally in this book, the essential message to come from these analyses is that the actualité of city-region change falls a long way short of ESDP aims. To be clear, this is not a commentary on the objectives of the ESDP, which raises different questions that require debate (e.g. Jensen and Richardson, 2004). This conclusion is also not about the uneven application of

ESDP principles in policy guidance across European space (Faludi, 2003). Rather, it is recognition that what is happening in villages and small towns in city hinterlands is not providing the kind of outcomes for environmental sustainability, landscape coherence, social inclusion or access to services that the ESDP promotes. Moreover, this outcome appears to be common for all nations, so it holds whether land-use planning regulations are tighter or looser, whether the countryside is much admired or still has a lingering negative image as a place of residence, and whatever the orientation of national political cultures and housing systems (Table 7.1). To achieve ESDP aims in other words requires much more than identifying the 'best' system or structure. It necessitates asking fundamental questions about shared orientations in dominant modes of behaviour and in core assumptions about what is appropriate behaviour in European societies. It might be that the answer is 'All about the people, stupid!'. Yet it is also 'all about priorities', assuming governments and citizens have a real commitment to the principles that lie at the heart of the ESDP.

## References

Ball, M. and Grilli, M. (1997) *Housing Markets and Economic Convergence in the European Union*, Royal Institution of Chartered Surveyors, London

Bennett, R.J., Graham, D.J. and Bratton, W.J.A. (1999) The location and concentration of businesses in Britain: business clusters, business services, market coverage and local economic development, *Transactions of the Institute of British Geographers*, 24, 393-420

Breheny, M. (2004) Sustainable settlements and jobs-housing balance, in H.W. Richardson and C-L.C. Bae (eds.) *Urban Sprawl in Western Europe and the United States*, Ashgate, Aldershot, 11-36

Commission of the European Communities (1999) *ESDP – European Spatial Development Perspective: Towards Balanced and Sustainable Development of the Territory of the European Union*, Office for Official Publications of the European Communities, Luxembourg

Countryside Agency (2004) *The State of the Countryside 2004*, Countryside Agency Publications, Wetherby

Countryside Commission (1997) *Public Attitudes to the Countryside*, Countryside Commission Postal Sales, Waldrave, Northampton

Diepen, A.M.L. van (2000) Trip making and urban density: comparing British and Dutch survey data, in G. de Roo and D. Miller (eds.) *Compact Cities and Sustainable Urban Development*, Ashgate, Aldershot, 251-259

East of England Regional Housing Forum (2003) *East of England Regional Housing Strategy 2003-2006*, Government Office for the East of England, Cambridge (this is available at http://www.go-east.gov.uk/About_Us/Business_Groups/Social_Inclusion/ Housing/Regional_Housing_Board/)

Esping-Andersen, G. (1990) *The Three Worlds of Welfare Capitalism*, Polity, Cambridge

Faludi, A. (2003) Unfinished business: European spatial planning in the 2000s, *Town Planning Review*, 74, 121-140

Faludi, A. and Waterhout, B. (2002) *The Making of the European Spatial Development Perspective: No Masterplan*, Routledge, London

Guy, C.M. (1991) Urban and rural contrasts in food prices and availability - a case study in Wales, *Journal of Rural Studies*, 7, 311-325

Harloe, M. (1995) *The People's Home? Social Rented Housing in Europe and America*, Blackwell, Oxford

Heitkamp, T. (2002) Motive und strukturen der Stadt-Umland-Wanderungen im interkommunalen Vergleich, *Forum Wohneigentum*, 1, 9-14

Henderson, S.R. and Hoggart, K. (2003) Ruralities and gender divisions of labour in Eastern England, *Sociologia Ruralis*, 43, 349-378

Hoggart, K. (2000) The changing composition of the rural population of England 1971-1991, in R. Cresser and S. Gleave (eds.) *Migration Within England and Wales Using the ONS Longitudinal Study*, The Stationery Office, London, 16-29

Hoggart, K. and Henderson, S.R. (2005) Excluding exceptions: housing non-affordability and the oppression of environmental sustainability?, *Journal of Rural Studies*, 21, 181-196

Hoggart, K., Buller, H. and Black, R. (1995) *Rural Europe: Identity and Change*, Arnold, London

Ismaier, F. (2002) Strukturen und motive der Stadt-Umland-Wanderung, in F. Schröter (ed.) *Städte im Spagat zwischen Wohnungsleerstand und Baulandmangel*, RaumPlanung spezial 4, 19-29

Jensen, O.B. and Richardson, T. (2004) *Making European Space: Mobility, Power and Territorial Identity*, Routledge, London

Kahn, M. (2001) Does sprawl reduce the black/white housing consumption gap?, *Housing Policy Debate*, 12, 77-86

Kemeny, J. (1995) *From Public Housing to the Social Market: Rental Policy Strategies in Comparative Perspective*, Routledge, London

Local Government Association (2002) *Key Workers and Affordable Housing*, London (also available at http://www.lga.gov.uk/documents/briefing/our_work/social%20affairs/affordablehousing.pdf

Mata Porras, M. (1995) Influence of the Common Agricultural Policy of the European Community on socio-economy and land degradation, in R. Fantechi, D. Peter, P. Balabinis and J.L. Rubio (eds.) *Desertification in a European Context*, Office for Official Publications of the European Communities, Luxembourg, 509-521

Morrison, N. (2003) Assessing the need for key-worker housing: a case study of Cambridge, *Town Planning Review*, 74, 281-300

Murdoch, J. (1995) Middle class territory? some remarks on the use of class analysis in rural studies, *Environment and Planning*, A27, 1213-1230

Norfolk County Council (1999) *Structure Plan*, Norfolk County Council, Norwich

Ostendorf, W. (2001) New towns and compact cities: urban planning in the Netherlands between state and market, in H. Andersson, G. Jorgensen, D. Joye and W. Ostendorf (eds.) *Change and Stability in Urban Europe: Form, Quality and Governance*, Ashgate, Aldershot, 177-192

Paniagua, A. (2001) European processes of environmentalization in agriculture: a view from Spain, in H. Buller and K. Hoggart (eds.) *Agricultural Transformation, Food and Environment*, Ashgate, Aldershot, 131-166

Paniagua, A. (2002) Counterurbanization and new social class in rural Spain: the environmental and rural dimension revisited, *Scottish Geographical Journal*, 118, 1-18

Richardson, T. (2000) Discourses of rurality in EU spatial policy: the European Spatial Development Perspective, *Sociologia Ruralis*, 40, 53-71

Tewdwr-Jones, M. and Williams, R.H. (2001) *The European Dimension of British Planning*, Spon, London

Tout, D. (1990) The horticulture industry of Almería Province, Spain, *Geographical Journal*, 156, 304-312

UK Department of the Environment, Transport and the Regions (2003) http://www.odpm.gov.uk/stellent/groups/odpm_control/documents/contentservertemplate/odpm_index.hcst?n=2263&l=2, originally released in 2000 as *Planning Policy Guidance Note 3: Housing*, The Stationery Office, London

UK Department of the Environment, Transport and the Regions (2003a) http://www.odpm.gov.uk/stellent/groups/odpm_control/documents/contentservertemplate/odpm_index.hcst?n=2263&l=2, originally released in 1994 as Planning Policy *Guidance Note 13: A Guide to Better Practice – Reducing the Need to Travel Through Land-Use and Transport Planning*, HMSO, London

# Index

Austria  25, 27, 157

Belgium  25, 27

Canada  76, 78
City-region competitiveness  9, 11, 12, 30, 75, 119, 135, 164
City-region delineation  6, 7, 28, 31, 42, 47, 97
City-region ecosystems  20, 36, 63, 64, 65, 87, 96, 124, 156
City-regions in peripheral regions  36, 44, 45, 56, 65, 71, 74, 75, 87
City-region policies  34, 37, 125-127, 129
City-region settlement structures
    Centralization of population and economic activities  14, 31, 44, 48, 63, 69, 71, 75, 76, 77, 119, 121, 122, 124, 133-134, 135, 139, 146, 147, 148, 158, 159, 165
    Compact settlement structures  11, 12, 65, 90, 124, 156, 166
    Decentralization of population and economic activities  6, 8-9, 23, 44, 47, 48, 49, 53, 55, 69-70, 71, 75, 76, 77, 96-97, 156, 158, 160, 166
    Dormitory / bedroom settlements  33, 55, 58, 59, 62, 83, 87, 105, 122, 125, 134, 135, 145, 147, 158, 160
    Rural hinterlands  2, 3, 9, 12, 13, 43, 45, 47, 49, 66, 72, 96, 100, 104, 119, 120, 127, 132, 155, 159, 160, 161, 164
    'Suburbanization'  5, 8, 9, 41, 43-44, 47, 48, 53, 64, 65, 126

City-region social forms  20, 34, 43, 55-56, 57, 114, 161
    Social exclusion  13, 124, 139-145, 158
    Social segregation / social exclusivity  12, 37, 66, 81-83, 91
    Social divergence/heterogeneity  34, 36, 70, 81, 82, 83, 103, 112, 157, 164
City-region sustainability  9, 10, 11, 20, 21, 53, 69, 70, 78, 120, 140, 141, 145, 146, 148, 156, 161, 162, 166
Commuting  7, 23, 28, 31, 42, 72, 77, 106, 131, 133, 146, 147, 161
    Functional urban areas  20, 22, 23, 24, 25, 26, 27, 28, 30
    Trans-border commuting  14, 162
    Travel to work areas  4, 6, 9, 23, 25
    Work-residence separation  36, 59, 60, 70, 92, 103, 104, 105, 124, 126, 130, 133, 156, 158, 161
Conservationist sentiments  10, 11, 42, 43
    Rural resistance to urbanism  4, 31, 54, 83, 84, 142
Core cities in city-regions  2, 6, 8, 9, 10, 14, 22, 23, 25, 37, 41, 48, 71, 74, 76, 80, 83, 87, 99, 125, 127, 130, 131, 132, 134, 138, 142-143, 146, 163, 165
    Urban distance decay impacts  6, 33, 34, 44, 45, 47, 52, 56, 59, 60, 62, 65, 81-82, 97, 106, 112, 115, 121

Urban spill-over effects 23, 80, 95, 96, 106, 114-115, 122, 126, 127, 144, 145
Cross-national comparisons 3, 7, 12, 13, 28, 31, 158, 160, 164, 166

Data sources 22, 30, 31
Denmark 25, 27, 157
Development plans/policies 12, 21, 34, 43, 48, 49, 63, 64, 124, 129-130, 133, 147, 148, 157
    Regulating city-region change 12, 49, 54, 84, 120, 121-124, 166
    Rural policy 11, 73, 140-141

Eastern Europe 12, 27
Economic activity sectors
    Agriculture 4, 36, 41, 43, 51, 53, 55, 60-62, 72, 88-90, 91, 96, 97, 102, 103, 104, 106, 108, 109, 110-112, 113, 114, 119, 125, 126, 141, 159, 160, 161, 162, 163
    Construction 102, 103, 107, 108, 126, 138, 144, 147
    High-technology activities 23, 36, 48, 70
    Manufacturing 36, 37, 48, 58, 59, 76, 77, 78, 79, 102, 108, 110, 115, 125, 126
    Natural resources / energy 11, 96, 102, 126, 166
    Sectoral policies 11, 155
    Warehousing / wholesaling 8, 78, 126, 132
Employment 13, 14
    City-region labour markets 23, 76, 105, 120, 160
    Rural employment 9, 35, 37, 43, 58, 59, 103, 108, 111, 119, 141, 145, 148
    Small town employment 9, 77, 78, 130-131, 132, 136, 143, 145, 146, 148

Urban employment 9, 10, 13, 14, 33-34, 35, 45, 46, 48, 49, 56, 71, 72, 76, 82, 104, 105, 125
ESPON 20, 23-25, 27, 28, 30
European Commission 1, 11, 14, 28, 156
European economic core 9, 10, 69
European Spatial Development Perspective (ESDP) 2, 9-10, 11, 12, 13, 14, 19, 20, 21, 34, 36, 41, 42, 64, 65, 66, 69, 70, 98, 105, 115, 119, 124, 134, 146, 148, 155, 156, 157, 158, 160, 164, 165, 166, 167

Finland 25, 27, 157
France 5, 9, 13, 14, 22, 23, 24, 25, 27, 28, 29, 30, 31, 32, 33, 34, 36, 43, 69-94, 157, 158, 159, 160, 161, 162, 163, 164, 165, 166

Gentrification 48, 53, 113, 132
Germany 5, 9, 13, 14, 22, 23, 25, 27, 28, 29, 30, 31, 32, 33, 36, 41-68, 157, 158, 159, 160, 161, 162, 163, 165
Globalization 7, 10, 23, 37, 101, 106, 111, 114, 115
Greece 25, 27, 28, 29, 30, 34, 36, 157

Heritage 11, 36, 43, 64, 124, 156
Housing 13, 14, 48, 54, 55, 75, 83, 86, 121, 133, 141, 144-145, 147, 148, 156, 157, 158, 159, 160, 162
    Gentrification 48, 53, 113, 132
    Housing affordability 37, 53, 55, 82, 95, 140, 141, 142, 144, 145, 146, 159, 161, 163
    Housing demand 14, 53-55, 72, 82, 83
    Housing growth 36, 37, 48, 102, 120, 122, 125, 128, 129, 132, 146, 147, 160

# Index

Housing / land costs  14, 37, 48, 51, 58, 65, 81, 82, 83, 88, 92, 114, 123, 139, 147, 157, 162
  'Planning profits'  50, 51, 52, 53, 57
  Second homes  13, 102, 107
  Social housing  37, 123, 141-142, 159, 162
Hungary  25, 27

Infrastructure  19, 37, 53, 58, 66, 69, 70, 78, 86, 99, 123, 131, 139, 163
  Transport and communications  69, 71, 92, 147
  Water and sewer systems  52, 58, 85, 86, 103
Ireland  27, 157
Italy  25, 27, 157

Land degradation  96, 161
Land ownership  51, 52, 54, 55, 84, 89-90, 111, 161
Land-use change  4, 5, 12, 13, 43, 119, 120, 146
Landscape change  5, 53, 54, 55, 61, 63
Local government  1, 22, 37, 47, 53, 66, 73, 77-79, 84, 88, 109, 125, 145, 163
  Inter-municipal competition  36, 49-50, 53, 57, 79
  Inter-municipal cooperation  32, 34, 70, 71, 72, 73, 74, 78, 84, 85-86, 120
  Local government finance  37, 50, 51, 52, 53, 57, 59, 84, 86, 103, 107, 109, 163
  Local political boundaries  1-2, 22, 42, 87, 125
Luxembourg  25, 27

Netherlands  12, 25, 27, 122, 156, 157
Northern Europe  10, 20, 27
Norway  25, 27

Peri-urban / peri-urbanization  3, 5, 6, 7, 20, 21, 22, 25, 28, 29, 30, 31, 32, 33, 34, 35, 36, 41, 42, 43, 47, 48, 51, 55, 61, 64, 65, 69, 75, 79, 81, 86, 87, 88, 90, 91, 92, 96
Polycentric development  2, 9, 12, 20, 23, 31, 91, 155, 156, 158, 164
Population change  2, 5, 14, 37, 45, 47, 48, 49, 50, 53, 74-75, 79, 80, 81, 82, 99, 100, 111, 125, 127, 128, 129, 134, 158
  Rural in-migration  19, 30, 37, 59, 66, 75, 83, 84, 91, 99, 102, 109, 111, 113, 123, 160, 161, 163, 164
  Rural out-migration  8, 36, 79, 95, 98, 102, 108
Population / housing density  4, 7, 8, 10, 19, 20, 25, 28, 29, 31, 42, 43, 47, 54, 65, 72, 144, 148, 164
Portugal  27, 157

Quality of life  2, 11, 13, 19, 45, 46, 47, 65, 87, 90, 91, 95, 103, 119, 124, 157, 163

Recreation / leisure  36, 48, 62, 96, 100, 104, 106, 107, 109, 113, 125, 132, 134, 136, 137, 138, 161
Regional economies  1, 2, 8, 20, 49, 110, 148
  Inter-regional inequalities  9, 10
  Networks  1, 12, 20, 23, 25, 78
  Regional government  1, 109
Rural areas
  As 'the good life'  2, 12, 113, 114, 122-123, 159, 161
  Definition / meaning  6, 25, 26-28, 29, 42, 72, 95
  Rural landscape  2, 5, 6, 65, 81, 87, 88-90, 92, 110, 113, 166

Rural lifestyles 5, 25, 34, 57, 70, 83, 88, 97, 164
Rural typologies 7, 21, 23, 28, 31, 32, 34-35, 36
Rural-urban
  Rural-urban duality 4, 5, 19, 21, 26-28, 42, 70, 72, 73, 98, 121, 135, 155
  Rural-urban fringe 3-4, 5, 7, 20-21, 34, 43, 44, 121, 129, 133, 147
  Rural-urban linkages 3, 6, 7, 8, 9, 12, 13, 20, 22, 25, 28, 30, 31, 33, 33, 34, 41, 43, 69, 75, 87, 98, 120, 133, 134
  Rural-urban interdependencies 11, 19, 23, 30, 34, 42, 70, 98, 119, 155
  Rural-urban partnerships 2, 10, 14-15, 19, 72, 155, 164

Service industries 14, 36, 37, 49, 53, 75, 76, 84, 91, 102, 103, 104, 105, 108, 110, 112, 115, 122, 125, 134, 135, 137, 138, 139, 140, 147, 158, 163, 164-166, 167
  Business services 10, 76, 125, 126
  Childcare services / kindergartens 57, 84, 165
  Cultural services 10, 11, 36, 43, 48, 64, 81, 87, 104, 109, 165
  Education services 78, 84, 91, 99, 104, 106, 107, 109, 122, 134
  Health services 78, 79, 99, 103, 104, 109, 165
  Library services 104, 106, 134, 135
  Retail services 76, 104, 122, 125, 131, 132
  Social services 78, 79, 84, 99
Slovak Republic 25, 27
Slovenia 25, 27

Small town – rural hinterland relationships 6, 10, 34, 97, 120, 123, 137, 144, 145, 148
Social change 2, 5, 11, 14, 21, 30, 61, 75, 81, 102, 103, 111, 115, 163, 164
  Social conflict 2, 96, 112, 163, 164
Southern Europe 8, 10, 20, 27, 98
Spain 5, 13, 14, 22, 25, 27, 28, 29, 30, 32, 34, 35, 36, 95-117, 157, 158, 159, 160, 161, 162, 163, 164, 165, 166
Spatial Planning 19, 20, 21, 31, 41, 78, 155
  Green belts 4, 121
  Industrial parks 37, 59, 79
  Land-use planning 4, 5, 11, 12, 13, 19, 43, 50-51, 52, 53, 55, 61, 63, 70, 73, 88, 90, 119, 120, 121, 124, 126, 132, 133, 138, 139, 141, 147, 156, 157, 160, 166, 167
Sweden 25, 27, 157
Switzerland 14, 25, 27, 70, 71, 77, 83, 162

Tourism 13, 14, 36, 37, 70, 82, 100, 101-102, 103, 104, 105-110, 11, 113, 114, 135, 160, 161, 162, 165
Transnational linkages 13, 14, 71, 162
Transport 11, 36, 37, 64, 69, 70, 72, 75, 80, 81, 82, 92, 96, 124, 125, 126, 129, 139, 143, 145, 148, 156, 160, 165, 166
  Public transport 12, 37, 49, 104, 106, 109, 135, 139, 140, 143, 145, 156, 165
  Traffic volumes 59, 63, 79, 131

United Kingdom 4, 16, 8, 9, 11, 13, 14, 25, 27, 28, 29, 30, 32, 35, 119-153, 156, 157, 158-159, 160, 162, 163, 165, 166

United States of America 3, 7, 8, 12, 76
Urban decline 30, 81, 96, 99, 127
Urban definitions 22, 23, 24, 25, 26-28, 29, 42, 72
Urban growth 8, 12, 14, 45, 51, 53, 80, 110, 111, 132
Urban hierarchy 10, 14, 22, 23, 49, 87, 131
Urban renewal 70, 120, 122, 124, 132, 134, 135, 146, 156

Urban sprawl 4, 5, 12, 19, 21, 43, 44, 48, 52, 64, 82, 96, 100, 135, 156
   Dispersed development 9, 12, 36, 44, 48, 53, 76, 77, 78, 97, 112, 114, 119, 159, 166
   Urban physical expansion 11, 19, 21, 22, 36, 42, 48, 53, 63, 66

Young people 37, 56, 83, 84, 125, 145, 161, 162

United States of America, 3, 7, 8, 12, 76
Urbanisation, 36, 54, 66, 69, 127
Urban dominance, 22-23, 24, 65, 120-28, 79, 81, 82, 89
Urban growth, 6-12, 16, 45, 51-53, 66, 110-11, 133
Urban hierarchy, 10, 11, 22, 25, 39, 47, 77
Urban renewal, 10, 120-22, 129, 131, 132, 134, 135, 136-150

Urban sprawl, 4, 5, 7, 10-12, 16, 18, 27, 36, 37, 66, 100, 123-25
Unplanned development, 9, 15, 20, 34, 48, 51, 79, 129-30, 115, 121, 119, 122, 106
Urban physical environment, 11, 19, 21, 22, 36, 42, 48, 52, 63, 66

Youngstocole, 1, 20, 85, 81, 117, 139, 145, 146-150

WITHDRAWN